T0323446

Borderland

JOURNALISM AND POLITICAL COMMUNICATION UNBOUND

Series editors: Daniel Kreiss, University of North Carolina at Chapel Hill, and Nikki Usher, University of Illinois at Urbana- Champaign

Journalism and Political Communication Unbound seeks to be a high-profile book series that reaches far beyond the academy to an interested public of policymakers, journalists, public intellectuals, and citizens eager to make sense of contemporary politics and media. "Unbound" in the series title has multiple meanings: It refers to the unbinding of borders between the fields of communication, political communication, and journalism, as well as related disciplines such as political science, sociology, and science and technology studies; it highlights the ways traditional frameworks for scholarship have disintegrated in the wake of changing digital technologies and new social, political, economic, and cultural dynamics; and it reflects the unbinding of media in a hybrid world of flows across mediums.

Other books in the series:

Journalism Research That Matters
Valérie Bélair-Gagnon and Nikki Usher

Reckoning: Journalism's Limits and Possibilities
Candis Callison and Mary Lynn Young

*News After Trump: Journalism's Crisis of Relevance
in a Changed Media Culture*
Matt Carlson, Sue Robinson, and Seth C. Lewis

*Democracy Lives in Darkness:
How and Why People Keep Their Politics a Secret*
Emily Van Duyn

Imagined Audiences: How Journalists Perceive and Pursue the Public
Jacob L. Nelson

*Pop Culture, Politics, and the News: Entertainment Journalism
in the Polarized Media Landscape*
Joel Penney

*Building Theory in Political Communication:
The Politics-Media-Politics Approach*
Gadi Wolfsfeld, Tamir Sheafer, and Scott Althaus

Borderland

Decolonizing the Words of War

CHRISANTHI GIOTIS

OXFORD
UNIVERSITY PRESS

Oxford University Press is a department of the University of Oxford. It furthers
the University's objective of excellence in research, scholarship, and education
by publishing worldwide. Oxford is a registered trade mark of Oxford University
Press in the UK and certain other countries.

Published in the United States of America by Oxford University Press
198 Madison Avenue, New York, NY 10016, United States of America.

Library of Congress Control Number: 2022941571

ISBN 978–0–19–756580–3 (pbk.)
ISBN 978–0–19–756579–7 (hbk.)

DOI: 10.1093/oso/9780197565797.001.0001

For Steph, a safe harbor, whose infinite welcome and respect
for alterity created so much beauty and fun in the world.
And for those who are coming.

Contents

Acknowledgments

No journalism is possible without interlocutors. The same is true for the progress of thoughts. In a work like this, where every moment of life becomes part of the research, even more so. I owe so many people so much.

First and foremost, thank you to the three most important points of my compass who always bring me home. To my truly wonderful and inspiring mother Nicky and brother Kosta, thank you for always loving and encouraging me and thank you for keeping me grounded. To my father Mick, thank you for our midnight talks on philosophy.

It is impossible to name them all, but some attempt must be made to acknowledge the friends, neighbors, teachers, colleagues—the village—who brought me to this place. In chronological order:

Moya Nowland, thank you for your kindness, for teaching me to read, for setting me on a path.

Mr. Symons and Mr. Graff, Rick and John, thank you for decolonizing my reading. For being outstanding teachers. Mr. Graff thank you for still supporting me! For your valuable edits and feedback on this work. I am unbelievably fortunate.

Chris Nash and Wendy Bacon, thank you for being the most inspiring journalism and theory mentors imaginable.

Steve Madgwick for loving me for 12 years and for loving the exploration of the world. For challenging, for expanding.

Devleena Ghosh and Keren Winterford, my PhD supervisors, this work owes so much to your generosity, to your intellectual gifts. Thank you, always.

Suz Flatt and Cale Bain, over the years of toil, your friendship toward me and support of this work went above and beyond. I know I will have a lifetime of wine and fun to try to repay.

Lucy Fiske, Sarah Phillips, Suda Perera, thank you for your crucial intellectual engagement and friendship.

Derek Wilding and Peter Fray, the opportunity you gave me to complete this work at the Centre for Media Transition was invaluable, your support incredible—thank you.

To my writing group, both amazing friends and rigorous colleagues, instigator Mehal Krayem, Judy Betts, Burcu Cevik-Compiegne, Irwin Compiegne, Bilquis Ghani, Helary Ngo, Kate Sands, and Cale Bain (again). You are the best.

Series editors Nikki Usher and Daniel Kreiss and Oxford University Press editor Angela Chnapko, thank you for your intellectual leadership in opening this interdisciplinary space. Thank you for guiding a first-time author. Nikki thank you for the intellectual curiosity, which allowed a young scholar to bend your ear at a conference.

To the anonymous reviewers: wow! what generosity and what skill and judgment. I am so grateful for your deep engagement. This work transformed with your input.

Finally, to the many journalism and research interviewees, whether your words made it onto the page or not. This book is only possible because of you. Thank you.

Introduction

A Global Polity

Dateline: Kosti, Sudan, July 29, 2011

Grim, solid, and determined, Helen was a hospital orderly; smiling, energetic, and fun, Emmanuel, a carpenter; curious teenagers, 13-year-old Esther and 14-year-old Sarah were hiding from the sun by lying under a bed, listening to a portable radio. These were a few of the people I met, who became my interviewees, on the banks of the Nile River at Kosti New Port. They were part of a mass of 20,000 people living in a makeshift refugee camp, made stateless by the Sudan, South Sudan split. Ethnically southern Sudanese, they had been living in the north; indeed some had been born there and never set foot in the south, yet when South Sudan was declared its own country they lost their Sudanese citizenship. They were now trying to get themselves, and their remaining possessions, to the new entity of South Sudan, where they could claim citizenship and once again be part of a national polity, and through that, once more be part of the global polity. The only realistic option was via river barge down the Nile—hence the aggregation of humanity, harboring both excitement and trepidation, at the port of Kosti.

They had been stuck there for months.

Unfortunately, the supposed peace deal between north and south had conveniently ignored the ongoing conflict in the oil-rich region on the border of the two countries. Any barge would have to travel across this disputed territory, through the resource-fueled war zone, and so all passenger transport was at a halt.

At that point, in late July 2011, the returnee camp was being hidden from the world. The international community (the United Nations International Organisation for Migration and the international nongovernment organizations facilitating the camp) deemed the situation too complex to explain, so journalists were not being alerted to the situation.

Borderland. Chrisanthi Giotis, Oxford University Press. © Oxford University Press 2022.
DOI: 10.1093/oso/9780197565797.003.0001

Twenty thousand lives simply put on hold—expelled from the global public sphere.

Tough luck.

I had come upon the crisis in Kosti by chance, five weeks into a typical, one might even say stereotypical, journalist's adventure. The plan was to spend six months traveling from Cairo to Cape Town by public transport, along the way finding and publishing the most inspiring stories possible. The goal of the independently run, UK-based website that had been created to publish the stories was to present a narrative different from the mainstream media's. A story about a refugee crisis was not on our news agenda. I first heard rumors about the situation in Khartoum, while trying to book a river barge passage myself, and to be honest I didn't believe the rumors. South Sudan's independence was a huge news story—how could this not be known?

Despite being arrested, having equipment confiscated, photos deleted, and being strongly advised by the police commander to leave the country as soon as possible, the stories from Kosti were published, once safely in Ethiopia, on July 29, 2011. Publishing was doing something, but not much changed.

The thousands of people of Kosti, stranded in a place with facilities set up for only 500, were stuck there until late May 2012, when an airlift operation began. With a 20-kilogram bag limit imposed, they lost their possessions, the last remnants of their old lives, with which they had hoped to start their new lives.

Fast-forward to the 2020s and South Sudan is still struggling with the effects of a drawn-out multifaceted conflict. A unity government was formed in 2020, but insecurity is constant. Almost one-third of the population has been uprooted. Two hundred thousand people are in internally displaced persons (IDP) camps. I don't know if the people I met in Kosti had their lives decimated again, or if they even had the opportunity to restart them.

This book documents the search for answers to questions that started to form in that camp in 2011. This time the borderland war focused on is the Democratic Republic of Congo, and this time I am working not just as a journalist but also as an academic. As I discovered in 2011, journalism is necessary but not enough. There are serious intellectual questions that must be asked of the role of journalists, and their audiences, in these borderlands and in these seemingly unending wars. And these questions are urgent.

The Refugee Is Not a Marginal Figure

Today, *every two seconds*, a person is displaced, caught in one of the more than 40 active conflicts around the world (Prigg 2017; Mohdin 2018). The vast majority of these conflicts have raged for longer than World War II and show no sign of ending. The war in the east of the Democratic Republic of Congo has been ongoing since 1994, uprooting millions of people and directly and indirectly resulting in the deaths of millions more. In the west, we are in a political era where our border policies are underpinned by unending wars that people choose to forget and mitigate rather than confront as the new reality (Duffield 2007).

At this critical juncture, what work are our daily storytellers engaged in to explain a complex interconnected world in transition? Are journalists, especially those engaged in foreign correspondence, making connections across time and space, across politics, economics, environments, and crucially, people? In the daily kaleidoscope of words, images, and sounds, many dispatches are outstanding, while many others are not. Many rework tired ideas and stereotypes obscuring the major dynamics of our time rather than elucidating them. The flaws are not random; they are the result of a profession born in the service of colonialism that has yet to decolonize, which, in fact, has yet to even acknowledge the need to decolonize. These flaws also result from a profession that operates under extreme risk and haste but has no fully developed philosophy of praxis. I propose a diagnosis and prescription in the hope of disrupting this state of affairs.

This interdisciplinary study argues that a decolonized practice of foreign correspondence would strengthen the potential for the realization of a global polity. It develops the existing critique of Anglo-American reporting on sub-Saharan Africa by focusing on the perceived and pronounced legacy of colonialist thinking. It argues that the recycling of old tropes and stereotypes obscures, instead of reveals, the postcolonial present. The specific words used to write about these places must change as part of decolonizing the profession. Less obviously, the way we conduct ourselves in borderlands is also key to this process. Borderlands are stratified so that professionals entering these spaces can end up interacting with locals in neocolonial ways. I argue this forestalls understanding of the local and global economics, politics, and discourses manifesting in new wars. It also stymies the potential for audience connection through storytelling; our practices as foreign correspondents incubate the words and ethics produced.

These arguments are made through a double-layered narrative. Two news feature articles that I wrote about, and from, the east of the Democratic Republic of Congo, as an Australian journalist with an Australian audience in mind, served as vehicles for reflection and also revision of practice. The DRC was chosen as the site of the journalism because it continues to be reported on, and understood, in a way that points to the persistent representational life of the extreme racial stereotypes sown in the colonial era. Examining the history of these stereotypes in detail allows us to see the colonial heritage of foreign correspondence and, from thence, to envisage and test new methods of meaning-making for journalism. A reflexive, autoethnographic essay documents my journey to Goma, DRC, and my personal responses to the events encountered. This narrative essay is overlaid with an analytical essay traversing multiple literatures on the challenge of meaning-making, as journalists, in postcolonial humanitarian and war zones.

Here I introduce the methodology and structure of the book. But first I discuss three interrelated, and foundational, issues that underpin the arguments of this research. Point one is on borderlands and new wars. War has shifted form dramatically and is leading to new configurations of human suffering. Our narratives have only just scratched the surface of recognizing this reality, and the concept of the borderland is not widely enough discussed. In this introduction I define what I mean by borderland.

The second issue discussed here is the question of structure and agency. An individual consciousness is required for decolonization work, but as professionals we exist within structures beyond our individual control. Moreover, our individual consciousness, which shapes our agency, is impacted by the structures, professional, personal, societal, that we inhabit. In this research, given the autoethnographic, practice-led nature of the method, it is especially important to acknowledge, up front, the personal-societal structures that have shaped the academic questions asked.

Third, the concept of a cosmopolitan global public sphere is discussed in relation to Ulrich Beck's work on "world risk society" (2009). In this reading cosmopolitan should not be understood in the common sense of a metropolitan elite lifestyle. It is the condition experienced by all of us now and especially by those whom the unequal currents of globalization impact most forcefully. The cosmopolitan condition is found in the borderlands created by new wars.

New Wars and Borderlands

New wars have been described as "privatized violence" (Beck 2009: 147) and wars of "persistence" (Kaldor 2016: 147). They are "a mutual enterprise in which the various armed groups have more to gain from war itself, from fighting, than from winning or losing, either for political reasons (to sustain and reinforce extremist ideologies) or for economic reasons (loot and pillage, or organised crime)" (Kaldor 2016: 147). The conflict in the eastern region of the DRC is emblematic of a persistent new war. It has been generating displaced people since 1994. As the site of the world's largest peacekeeping mission, the erstwhile town, now city, of Goma, attracts legions of UN and development workers. Charged with making peace in the midst of an unending war, they have in fact created a world of international interveners that disrupts and recreates the local space. This is what international relations scholar Séverine Autesserre (2014) calls "Peaceland" and is similar to the concept of "Aidland" used by critical scholars of development. This understanding is important if we are to operate reflexively in these places.

Goma is also a borderland. Goma exists on the edge of two countries; the borderland referred to is real in its effects of war and on life trajectories, and in the way the social geography changes dramatically at the border where Goma turns into Gisyeni. Yet the concept of a borderland is also used to interrogate what is meant by borders in our globalized world (Agier 2016). The issues in this book are not a problem of "over there" but of everywhere. The borders between plenty and poverty, citizenship and asylum, care and cruelty surround us. Myambo and Frassinelli (2019) remind us that borders take all forms, especially in cities that are strongly acted upon by global forces. Myambo deploys the concept of cultural time zones (CTZs) to discuss how cities have been "reshaped by globalization into complex, highly-texturized, immensely-variegated urbanscapes of divergent spatio-temporal microspaces . . . globalizing cities have created harder and harder borders between the CTZs frequented by the haves and the have-nots" (Myambo and Frassinelli 2019: 279). Conflict does not shield a city like Goma from these sorts of social and cultural boundaries—the conflict in fact globalizes the city and creates the boundaries. The border dynamic of "selective inclusion and exclusion of human beings based on their categorization of more or less desirable" (Myambo and Frassinelli 2019: 283) operates at the club door (in New York and in Goma) and at the walls of Europe and America—and both are the result of ongoing conflicts. The middle and upper classes of globalized cities retreat behind CTZ enclaves in fear of the increasing inequality of the world. The same is true for middle- and upper-income countries that excise parts of their territories, creating larger and larger border zones that become "key sites in the transnational government of mobile bodies and the management of populations" (Aradau and Tazzioli 2019: 200). The transnational governance that begins in the Peaceland of Goma extends to the shores of Europe or Australia, and this global dynamic then links to the widespread desire for deglobalization visible in increasingly xenophobic border politics (Myambo and Frassinelli 2019).

While borders and borderlands are spaces of international and transnational governance, they are also spaces of humanity. Anthropologist Michel Agier analyzes the borderlands of Europe and the Americas as well as African conflict zones and refugee camps as spaces of dynamism creating new moments of social and identity potential through local-global cultures (Agier 2016). He argues the importance of understanding the political and human dimension of the border as one of the key contemporary dynamics of our time, and to do this, "It is necessary to expand considerably the sample of places and moments taken into account" so as to include what Agier calls "border situations" (2016: 17).

In dealing with the issues raised by these border situations, and as professionals who can (relatively) easily cross borderlands—in fact are expected to—the key question is, do foreign correspondents have the potential to be conduits for new meaning-making in the context of an embryonic global public sphere?

On the surface the obvious answer is yes. The press corps exists to "serve as the extended eyes and ears of the public" (Carey 1987: 5). A crucial role *could be* played by journalists in creating the sociological imagination needed to foster a global public sphere. However, in 2002, commenting on the lack of sociological imagination often visible in the war reporting of the 1990s, Greg McLaughlin argued that there are "many external and internal forces corroding what imaginative potential might exist in mainstream journalism to make sense of war" (McLaughlin 2002: 4). This critique still rings true, and there is more to say. Today, foreign correspondents must go further. To make sense of unending wars they must connect our fragmented understanding of the world we are all, synchronously, creating. In her 2014 book, which sought to draw attention to the scale of expulsions currently taking place (not just of people, and not just in the developing world), sociologist Saskia Sassen argued that new geographies of extreme zones for key economic operations are materializing before our eyes, but they do not receive enough attention, and neither is their significance understood. At one end of these dynamics are the finance centers of global cities, and at the other are outsourcing zones (2014: 9). Sydney, for example, is one such minor global city, one of the most expensive in the world to live in, fueled in part by financiers and lawyers. Goma is also a global city in that it is an outsourcing zone. Its population is deeply embedded in the movement of raw resources supplying high-tech industries, legally and illegally, and as part of the growing reach of shadow economies.

Sydney and Goma are materially connected through the complex and often random assemblages of global economics. However, at the same time, as part of a more systemic global dynamic based on discourse, the people within these cities are being actively disconnected. If eminent security and development theorist Mark Duffield (2007) is correct, we live in an era of unending war and also in an era where humans are being divided between insured and uninsured lives. Duffield's argument holds that in a city such as Sydney a substantial majority of the working, middle, and upper class exist as developed "species-life" (2007: ix): "Developed life is supported and compensated through a range of social and private insurance-based benefits and bureaucracies covering birth, sickness, education, employment and pensions. In contrast, the underdeveloped or 'non-insured' life existing beyond these welfare technologies is expected to be self-reliant" (Duffield 2007: ix). Duffield makes the point that the uninsured human, for example, the refugee, is disciplined to be self-reliant through the technologies of development. Thus, if we apply Duffield's argument, walking into an Oxfam shop in Sydney and giving a goat for Christmas is actually "central to the new or culturally coded racism that emerged with decolonization" (2007: ix), which sees self-reliant life as good enough for some and supported life (including support to shop excessively for Christmas) as the bare minimum for others. Development, says Duffield, is not about "narrowing the life chance gulf"; it is a technology of security, attempting to hold the victims of war and poverty in situ (2007: 19).

In short, important currents of our times, regarding conflict and development, are either hidden or dressed up to appear the opposite of what they are. Our narratives must shift if we are to understand the processes by which we betray the other—and the processes by which we don't.

In this endeavor of personal and structural decolonization, generations of people from all over the globe have been helped by Franz Fanon's words on the impact of colonization on the colonized. For me too, Fanon was my starting point . . .

Dateline: Sydney, Australia, March 23, 2013

It wasn't my first time reading *The Wretched of the Earth*, but this was a new edition with a foreword by Homi Bhabha titled "Framing Fanon." Nine years earlier, when this edition had been released, I had little time for reading; I was already working crazy hours at a local newspaper. In my deadline-driven pursuit of "the news of the day," reading anything other than council reports and the words of other journalists, went out the window. To cap off my unreflexive practice, I almost killed myself driving in haste to get to stories—not once but twice.

This edition told something of the history of the book, of its effect on independence movements around the world. The Black Panthers, Steve Biko, H-Block of Belfast prison (where multiple copies were kept), the People's Mujahideen. I remembered its effect on me the first time I had read it. But now I felt something more; I felt I was holding a dynamo in my hands, and I felt a connection through space and time to these unknown others: "the dispossessed subjects of globalization . . . the wretched of the earth, in our time and Fanon's" (Bhabha 2004: xxviii). I was more than a year and 12,700 kilometers away from the moment when I would meet Furaha, a teenager, an orphan, a refugee, and a microentrepreneur, both an extreme victim and extreme self-sufficient survivor, who would become the key semiotic building block of one of the feature articles produced. As I read Fanon's words a journey had begun toward that meeting—but would we ever be really connected? At that moment I was lying on lush grass on a sunny Saturday, in a rate-payer-funded municipal park, where every Australia Day, the highlight of the council-run program is the citizenship ceremony. Next to me, also reading, I had my partner of seven years, with whom I had traveled the world and with whom I had lived in Europe (thanks to my luxurious holding of two passports—Australian and Greek).

While I lazed in that park, warm and content and doing exactly what I wanted to do, Furaha was in a refugee camp, labeled an internally displaced person. Her childhood had already included running for her life from distressed and drugged members of avaricious militias, watching the murder of her neighbors, her friends, her family.

Yet, she was no agency-less victim, no voiceless, blank face on a development NGO's advertisement. By the time I met her, she had built a life with the tools she had, showing amazing resilience and nous, despite the horrendous situation she had been put in. She was exactly the person Fanon addressed with such rage and reason.

When I lay there reading, I didn't know I would be going to the Democratic Republic of Congo. All I knew was that I would be going to refugee camps in sub-Saharan Africa and I would not be portraying the people I met as helpless victims. There had been too much of that already. That much I had known since I was a teenager myself, when I'd had the good fortune to pick up, again thanks to my fantastic, rate-payer-funded, local library, the book *Africa: Dispatches from a Fragile Continent*. Written by former *Washington Post* correspondent Blaine Harden, it was a remarkable book in many ways, not least because Harden conveyed that he was trying to correct the misrepresentation of the continent of Africa that the craft of journalism—including his own work—had created. Harden's was the first but not the last book I read from foreign correspondents stationed in sub-Saharan Africa that included an element of mea culpa. Many foreign correspondents, when they write of their time in Africa, away from the heat of news deadlines, seek to engage with, and to some extent correct, the common media representations that they and/or their colleagues and/or their news organizations have helped to create (Harden 1991; Dowden 2009; Sara 2007; Hunter-Gault 2006; Keane 2004; Zachary 2012; French 2005).

The late, influential academic Stuart Hall described the work of representation as the process of "classification, conceptualization and value making." But we also come to give meaning to "objects, people and events . . . by the frameworks of interpretation which we bring to them" (2013a: xix). This means that the process is circular, for those frameworks of interpretation are influenced by the value-laden representations we have already seen of those, or similar, objects, people, and events. Journalism is often described as the first rough draft of history. The description gives emphasis to the fact that some details of that early interpretation may be incorrect. Also highlighted in that description, whatever may be the corrections that come later, is that those early inscriptions of values are crucial to meaning-making.

So, it's that easy: changing the representation of sub-Saharan Africa is simply a matter of writing different stories so we can assign different values.

Except that it's not.

Journalism is not fiction; it is not *just* a process of putting words on the page, a process controlled by the author. It is a particular research practice designed to discover and narrate new information in, and about, specific times and spaces (Nash 2013). I wanted to present Furaha as an individual to understand her on her own terms but, at the end of the day, she was a young, female refugee in a camp in sub-Saharan Africa. I struggled to find the words with which to escape the political discourses she'd been placed in.

In my attempt to avoid the myth of the helpless victim waiting for western charity, I instead placed her in the category of the hard-working female microentrepreneur, and thus within the dichotomies of "hard-working versus lazy," "worthy versus unworthy," "female African (good) versus male African (bad)," and "good refugee versus bad refugee." In the feature article Furaha is a good displaced person, worthy of our admiration, or to put it in the terminology from whence these thoughts come, she is part of the "deserving poor"—a medieval "good pauper" (Mahon 1992: 36) because she's a female microentrepreneur, in her own country, and that's the lot in life that our international development frameworks, and our border policies, have allocated to her (Duffield 2007).

I knew that was what I was doing, but wherever I turned, I came up against some sort of western-imposed stereotype, some less-than-ideal representation. Just by being there, in a refugee/IDP camp, she was located in a discourse, she was fixed in a place that represented the start of potential migration to the west, and she was therefore seen in light of our own overarching frameworks of concerns, including our policies of immigration and border protection.

Feminist development scholars Maria Eriksson Baaz and Maria Stern, experts in the DRC, had a similar problem when they tried to research perpetrators, not victims, of war-crime rape. They found they didn't have "a lexicon for properly hearing and writing about rape differently—in a way that did justice to the stories the soldiers told" (2013: 7). I struggled to break out of the good/bad refugee dichotomy in part because of its strength, and in part because I had never been exposed to much writing that broke out of the grid of intelligibility created by the discourse of development; I didn't have a lexicon with which to write something else about an amazing young woman who found herself in a refugee camp. Our frameworks of understanding are not fit for purpose in part because they have not been decolonized and in part because that lack of lexicon does not impact most creators of symbolic reality; we are not the wretched of the earth . . .

Dateline: Harare/Hollywood/Sydney, July 28, 2015

The "world" is mourning Cecil the Lion. Thirteen years old he was photographed innumerable times by western, mainly British, tourists in Zimbabwe's Hwange National Park. There are calls for the US hunter who shot him with a bow and arrow to be prosecuted. Cecil's death will eventually result in changes to US law.

While some are mourning this animal's life others are drawing attention to the power relations revealed by this global story. My Instagram feed from a fashion house called PsychoSisters Hollywood has an image of an #AllLivesMatter protester in a Wizard of Oz lion costume. The post draws attention to the warped values evident in western media and political discourses, where animals originating from Africa are afforded respect and a privileged status denied to the continent's human inhabitants and descendants. The headline attached to the photo reads, "Going to start dressing like a lion. That way cops know that if they kill me. White people will avenge me."

Why do I follow PsychoSisters Hollywood on Instagram? Answer: I'm attracted to their rage. As discussed by Barrington Moore (1972), scholars cannot divorce themselves from their subject positions—particularly when engaging in social issues.

> For the student of human society, it is a truism that his conscious and unconscious moral assumptions and preferences will guide much of his work no matter how objective he tries to be. At the very least, a scholar quite satisfied with the world in which he lives is unlikely to start out by asking the same questions as the one who feels a sense of outrage and nausea upon picking up the daily newspaper. (Moore 1972: 4)

This necessity to acknowledge subject positions becomes more necessary for those scholars engaging in autoethnography. The questions I ask as a scholar, and the way I respond to the world, are influenced by my experiences. Experiences like the fact that a combination of events, including a glut in drug production in the "Golden Triangle" of Southeast Asia, led to a period of heroin-driven gang violence at my high school—a school at that time in the bottom 10 percent of the state in terms of socioeconomic disadvantage. I learned about the production glut because I was the "teenage rep" on a small police and community working group trying to figure out a strategy to deal with the fallout. It was an early lesson in the fact that the world is one global economic sphere and certain human beings get the pointy end of the stick. The fact that these drugs were ending up in our community was not random, and yet it was our community that was being demonized—as if it was our choice.

That early experience taught me that when dealing with the colonial and postcolonial project one cannot, for one minute, forget the interconnected economic and cultural power relations at play. I am also influenced by the belief that just because someone, or some many, are acted upon by power does not mean that they do not know exactly how they are being screwed over. PsychoSisters Hollywood do not need to have read V. Y. Mudimbe on the importance of exposing the knowledge systems created by the "colonial library" (1994: 213), or Achille Mbembe's identification of "beasts" as the "metatext" through which "discourse on Africa is almost always deployed" (2001: 1), or Raoul J Granqvist on the way descriptions of East Africa as an "Animal Eden" was used by *Life* magazine to obscure and trivialize anticolonial currents (2012). They may indeed have read these works, but even without them, their experiences have equipped them to utterly understand how unequal power dynamics still work to see African animals valued above black human beings. And they engage in tactics, operating with the tools they have.

The questions I ask are influenced by the fact that global spaces of disadvantage are also, more often than not, spaces of diversity from which new understandings of the world are born. As described by Agier in his anthropological work on borderlands:

> Cosmopolitanism is the experience of those women and men who experience the concreteness and roughness of the world . . . [W]ho better than the "uprooted" to give us the concrete and empirical trace of this new cosmopolitan condition, and to reflect on the political perspective that it establishes on a common world scale? (Agier 2016: 76)

I went to a primary school where our different cultures were *forcibly* celebrated, where we were forced to learn our parents' mother tongues. I had the luck of having an intelligent, caring, proud, male Indigenous teacher in Year

5—an early socializing experience that meant no stereotypes of Australia's first people could ever find purchase in my primary frameworks of understanding. The luck of attending a high school that traded on our diversity with the slogan "Friends from 54 nations," where Achebe's *Things Fall Apart* was taught instead of Conrad's *Heart of Darkness*. My world was full of power and the beautiful.

Right-wing MP Pauline Hanson was elected to Australia's parliament in the same year as the neoconservative John Howard began his long reign as our prime minister. Hanson's xenophobic rhetoric against Asian immigrants was given legitimacy and changed the political culture in Australia. Very early on during this change I vividly remember my friends of Vietnamese heritage sitting outside class in the maths block hallway, despairing. One close friend, who'd spent years in a horrible Malaysian refugee camp, where one of her earliest memories was being chased by a man with a knife, a friend who had arrived by boat, almost crying, asking, 'Why do they hate us, Chrisanthi?' We were 16 years old. Pauline Hanson was a painful shock. And yet I couldn't discard everything I had been taught so far. I held on to my mantra about multicultural Australia. Every culture equal, special, important.

More than two decades later I would sit next to a despairing friend again; this friend had arrived in Australia as a refugee from Afghanistan. Four days earlier, round metal spheres, turned into deadly force by human ingenuity and human hatred, were propelled into mums, dads, children, brothers, sisters, friends, neighbors, strangers—fellow human beings. The line from hateful words to hateful acts was completed.

The specific attack I refer to is Christchurch, New Zealand, March 15, 2018. The specifics of each attack matter. The lives taken matter. As newspapers filled with photos and stories of love and grief, the recognition of these everyday lives, tragically, made extraordinary through their death, was a crucial acknowledgment of our shared humanity—the moment of nonbetrayal that comes after the betrayal.

But there were other responses too. In the aftermath of the attack a right-wing politician in Australia who, thanks to quirks in the Australian electoral process, was voted into a senate seat with only 19 primary votes, used taxpayer money to release a media statement. He chose to describe violent white supremacism by an Australian, influenced by xenophobic politics at home, and extremist groups in Norway, as an unsurprising response to New Zealand's immigration intake, a program that had recently included high-profile support for refugees.

The Christchurch attack, in its globally fused xenophobic rejection of the other, is intimately linked to the themes in this book. Rewind to September 11, 2001, another horrific day of death that scarred families and society. Also the day the visible and invisible border defenses of the world's last remaining superpower were penetrated. The "global war on terror" entered the lexicon.

In the gap-filled narratives that emerged at the same time as the "global war on terror," somehow terrorist and refugee were inflated. Too often darkness and obfuscation pose as knowledge. I refer to this in chapter 1 as the "lacuna effect." This effect is perhaps most visible, and thus available for analysis, in the Democratic Republic of Congo because there the forces of the postcolonial world coalesce; in extremes you can see effects most clearly. But the key learning we take away is not limited to a geographically contained space. Versions of missing and misconstrued knowledge, where the interests of the demos are excluded, exist far too prevalently for a functioning global public sphere to operate. The task of building that global public sphere is urgent.

Defining a Global Public Sphere

Ulrich Beck in his book *World at Risk* (2009) argues we are living in an era of global crises that transcend national boundaries and national governments and that this creates a world risk society. In some respects this argument is self-evident, but it is also revolutionary. If this argument is correct, there are far-reaching, potentially extremely bleak consequences. The first point to unpack concerns the nature of risk itself, which is never a real thing because it is always a future event. In this way the decision to classify an event as a risk becomes all-important. Beck argues: "What 'relations of production' in capitalist society represented for Karl Marx, 'relations of definition' represent for risk society. Both concern relations of domination" (Beck 2009: 31–32). Brought into the context of one shared world of transnational risks, this means actions based on a dystopian "discourse of globalisation and managing its associated risks" (Heng 2006: 74) colonize the future for certain populations "who are (or are made into) 'risk persons' or 'risk groups' [thus] nonpersons whose basic rights are threatened" (Beck 2009: 16). The inequality dynamic at the heart of a world governed by risk applies equally to the unequal distribution of climate change effects, where the socially vulnerable are impacted first, and to the transformation of war. In war decisions today, "The emphasis is more on *averting* speculative scenarios rather than

attaining specific outcomes" (Heng 2006: 74) and this combines with new remote control war techniques that mean that civilian casualties rise. Civilians
in some (postcolonial) countries are at greater risk of suffering so that
civilians in other (western/northern) countries are (in theory) at less risk. In
short, "Risk divides, excludes and stigmatizes" (Beck 2009: 16).

The dynamic of social inequality is inherent in all politics around the concept of risk, whether at individual-to-individual, local, national, or global
level. However, paradoxically, when risk analysis is applied to a shared world,
it also unleashes a "cosmopolitan moment" that has the potential to overcome the division born in risk.

> The "cosmopolitan moment" of world risk society means, first, the *conditio
> humana* of the irreversible non-excludability of those who are culturally dif-
> ferent. We are all trapped in a shared global space of threats—without exit.
> This may inspire highly conflicting responses, to which renationalization,
> xenophobia, etc., also belong. *One* of them incorporates the *recognition* of
> others as equal *and* different, namely, normative cosmopolitanism. . . .
>
> From this perspective, the "cosmopolitan moment" has both a descrip-
> tive and a normative meaning. In the first place it is a matter of grasping the
> *reality* of non-excludable plurality which is driving the dynamic of world
> risk society, regardless of whether this reality is ignored and demonized or
> embraced and transformed into active global policy. Thus at the same time
> clues and principles become visible concerning which kind of cosmopol-
> itan thought and action is possible and could and should be realized. (Beck
> 2009: 56–57)

In this exhortation to recognize the reality of a shared global present there is
a clear role for journalism, in all its formats, to once again provide us with an
imagined community, as it did with the birth of the nation-state (Anderson
1983). To some extent this has already occurred; however, any sense of an
imagined global community has been found predominantly in elite institutional governance structures such as the United Nations, the International
Monetary Fund, and the Davos World Forum (albeit the latter also includes
some elements of grassroots globalization as counterprotest). This elite and
technocratic vision mirrors the problems with risk society where the power
to define rests with those endowed with the label of expert. Such a state of
elite and institution-driven global imaginary cannot continue if a global
public sphere is ever to reveal itself, for, as Beck makes clear, "A global public

discourse does *not* arise out of a consensus on decisions, but rather out of *disagreement* over the *consequences* of decisions" (Beck 2009: 59). Again there is a clear role for journalism here in fostering that global public debate which is not limited to experts but incorporates the *publics* who journalists aim to serve. Moreover, Beck goes on to warn:

> This is also decisive for the future of democracy: Do we depend on experts and counter-experts in all details of life-and-death issues—from our everyday lives to global transformations—or are we regaining the capacity for independent judgement through a visibility of threats that must be culturally created? (Beck 2009: 45)

Journalism scholars have not ignored Beck's warning. Simon Cottle (2009b) has drawn our attention to Beck's important sociological thesis and its implications for news media as emissaries of a global public sphere. He and his collaborators (Cottle and Rai 2006; Cottle 2009a, 2008, 2013) have also gone a long way in showing the importance of different global crises as reporting vehicles that provide a basis for developing a sense of a global public sphere. Moreover, different reporting frames have the potential for either activating or deactivating a sense of normative cosmopolitanism—that is, cosmopolitanism that recognizes different lives as equal and important. Cottle argues:

> How different global crises become communicated, contested and *constituted* within the world's media formations and communication flows—within its evolving global news ecology—may not only be the harbinger of a new (forced) cosmopolitanism but also prompt the *re-imagining of the political* within an increasingly interconnected, interdependent and *crisis-ridden* world. (2009b: 170)

Peter Berglez has likewise drawn on Beck to argue the need for a global outlook in all forms of reporting as a necessary precondition to a global political culture. He argues:

> Long-term goals such as a "world cosmopolitanism" or "global democracy"—whatever they might mean—*first of all*—require a journalism that increasingly engages in internalizing globalization and global relations in everyday news reporting. (Berglez 2013: 13)

Berglez agrees with Anderson's (1983) groundbreaking thesis on the role of journalism (and other writing products—like novels) on the production of the imagined community that had to be created for the modern nation-state to exist politically. He argues that not enough scholarly attention has been given to the idea that

> mainstream news media's ability to play a democratic role in society increasingly *presupposes* the ability to interconnect the world's continents (i.e., it is not enough to report events from distant countries with the rationale of foreign correspondence). (Berglez 2013: 13–14)

Berglez (2013: 15) goes on to argue the need for a new epistemological approach to global journalism. I agree on the need for a new epistemological approach but differ from Berglez in that I am not rejecting the role and rationale of foreign correspondence—instead I seek to adapt it. The development of the global public sphere can be incorporated into the professional raison d'être of foreign correspondence—especially when reporting from borderlands.

Borderlands are particularly important to this endeavor because the border gives the illusion that noncosmopolitanism is still an option. Borderlands and border zones delay the recognition that the other no longer exists (Beck 2009: 37) in the same way that the definition of human rights issues as "foreign policy" yet asylum seekers arriving in country as a "domestic crisis" (Cottle 2009b: 103–4) delays recognition of the interconnected nature of global risk inequality and the refugee crisis. Foreign correspondence can play a significant role here in collapsing the distinction between here and there, between self and other. However, this is not a given; in fact foreign correspondence has an opposite, colonial history. One of the first famous foreign correspondents was Henry Morton Stanley. His trip through the Congo was paid for by the Belgian king Leopold, and the quid pro quo was that Stanley would use his power of words to define the people of the Congo as in need of colonization (Dunn 2003: 24).

With such a history one wonders whether the profession should be rejected all together. The argument against such a rejection is that that the base principle of journalism is sound in terms of relational ethics. Political philosophers Arnsperger and Varoufakis (2003) argue social solidarity is formed through identification with the condition of unknown others—this is exactly what journalism, including foreign correspondence, can, should, and often does do. It takes stories of a stranger's condition in life and connects you to it. A global communicative public sphere built on the principles of normative cosmopolitanism is possible. Foreign correspondence, *because* of the way it incorporates relationships with "the Other," is crucial to that endeavor.

In this book, by moving discussion of a global public sphere to a borderland case study, three key strands of the emerging epistemology for global journalism are developed. First, applying a decolonizing lens to news frames allows us to see the continuing echoes of colonization that stymie the potential for a global public sphere. This then reveals the need for new genres and forms of knowledge creation that translate into new writing and reporting techniques. Global journalism needs to find a lexicon that matches current postcolonial realities and, at the same time, empowers human connection.

Second, it becomes clear that the reporting outlook required is not only global but in fact *glocal*. The social geography of borderlands highlights the dynamic glocal processes that need to be understood in our connected word of risk. Studying borderlands also highlights how professional relations are embedded in social relations and that current routines and practices of journalism can work to obscure those glocal realities. Reversing the current situation, correspondence from borderlands must seek out and focus on the local-global dynamic. Generating and narrating that knowledge is foundational to developing an epistemology of global journalism; knowledge about the world must be understood to involve knowledge of local-global processes.

Finally, a case study approach that uses autoethnography to follow the creation of specific pieces of journalism requires us to move beyond audience reception. So far scholarship has focused on the "relational ethics professionally embedded within news texts" (Cottle 2009b: 145) primarily at the level of semiotics (Chouliaraki 2006; Moeller 2006) or at the level of communicative architecture and newsroom decisions (Cottle and Rai 2006; Cottle 2009a, 2013; Berglez 2013). My focus opens discussion on the relational ethical praxis, on the ground, in crisis zones, that is necessary to create those news texts in the first place. Using the autoethnographic method reveals that professional ethics are grounded in a myriad of small and large decisions, that borderlands are especially ethically fraught places, and that philosophizing the most basic of reporting actions is a necessity, so as to strengthen the chances of ethical decision-making in the most complex of situations. Ethical praxis strengthens the knowledge claims of the profession; it is also absolutely necessary if foreign correspondence is to justify its continuance.

I will use the term "foreign correspondence" to describe reportage that mimics the practice of being a foreign correspondent. This means the correspondence may or may not have been created by journalists who are permanently stationed in a region or country, but they are journalists who have been trained, either through tertiary degrees or through on-the-job socialization, in the Anglo-American journalism tradition (wherever in the world that training may have taken place), and who report for news organizations whose headquarters are based in a major metropolitan city. This places my book within scholarship on "the Anglo-American ethnocentrism that pervades the study of mass media journalism and mass communication" (Harris and Williams 2018: x). However, as noted by Palmer (2018: 145), staff correspondents may be a dying breed, but they are still the "gold standard" of the profession. Exactly because it is still the gold standard, it is the practice based on traditional Anglophone correspondence that needs to be decolonized. Often, particularly in postcolonial contexts of ongoing border conflicts, it can be hard to distinguish where foreign correspondence begins and war correspondence ends. To that effect the literature on war correspondents is also relevant, and the terms may be used interchangeably. The dangers faced in foreign/war correspondence are changing form, and without decolonization this will impact frames and ethical relations of reportage; the increased uncertainty has the potential to cement existing bad practices and create new ones.

The Methodology and Structure of This Book

The overarching methodology of this work is practice-led research driven by the creation of the feature articles described in chapter 2. This creation of journalism artifacts produced "praxical knowledge . . . the particular form of knowledge that arises from our handling of materials and processes" (Bolt quoted in Smith and Dean 2009: 6). Accordingly, each chapter in this book employs different research methods and different theoretical coordinates. These decisions emerged based on the needs of the creative praxis and the data gained through the ethnographic and autoethnographic observations. Theory and method became an iterative web as I tried new journalism research methods, and learned more through new theoretical coordinates, which then led to me revisiting my reporting experiences and data with new eyes. For Donald Schön this sort of practice-led research is particularly suited for situations of uncertainty, which he argues are best dealt with in professions via a reflective conversation with the materials at hand (Schön 1989: @27m02s). This reflective conversation is needed because, by definition, a situation of uncertainty requires you to frame the problem before you know what questions to ask.

The meditation on the question of whether, and how, foreign correspondence can be decolonized and help build a global public sphere begins with postcolonial literature, traverses humanitarianism, peacekeeping, and development texts, and ends with the philosophy of Emmanuel Levinas. I use the term "meditation" consciously. The question took years to answer in conversation with journalism praxis and scholarship. Key moments of realization are scattered throughout this book, as datelines, and are part of the presentation of that data—they represent the process of reflection in, and on, action (Schön 1983). In presenting them in this way, I hope to elucidate the argument that ethical meaning-making is a process not simply of conscious representation but also of reflexive praxis. We live in worlds of words, and all work requires self-reflection so as not to reproduce the very problem one is trying to eliminate.

Necessarily, in that reflective conversation with materials, the researchers themselves become part of the process. This leads to the autoethnographic method and the honest admitting of mistakes that are part of this research. Michael Patton, in his authoritative textbook on qualitative research, argues that "creative fieldwork means using every part of oneself to experience and understand what is happening. Creative insights come from being directly involved in the setting being studied" (2002: 302). I kept ethnographic and autoethnographic field notes, a public-facing blog, and an exegesis diary where I was able to engage in "critical self-reflection . . .b[which involves] a willingness to consider how who one is affects what one is able to observe, hear and understand in the field" (Patton 2002: 299).

Engaging in ethnographic interviews around journalistic interviews and engaging in autoethnography while engaging in journalism was challenging. Taking notes on me interviewing, while trying to take down my interviewees' quotes, was a mind-bending task. The shorthand scribbles on both sides of the dividing line of my notepad have gaps, and I am not sure that particular activity is one I will attempt again. Furthermore, a limit of this practice-led, autoethnographic method is the small data sample. Accordingly, engagement with secondary literature is crucial for several of the arguments made throughout the book, particularly in chapters 3, 4, and 5. However, conversely and relatedly, a strength of the method is being able to draw lateral connections across data, as every element of praxis is part of the research.

The method adopted meant that professional and social structures, as well as personal agency, are constantly under interrogation, revealing tensions that can yield insight into the operation of the field. In taking this approach I am influenced by Pierre Bourdieu, the sociologist who has done the most to help us conceptualize the way in which social structures act upon us at an individual professional level. His theory of habitus argues that all people have a strategy-generating principle (habitus) that enables them to cope with unforeseen and ever-changing situations (Garnham and Williams 1980: 212). The habitus integrates early socialization from the home, previous professional socialization, and your interpretation of the rules of the game of the field in which you are currently operating. It is "a system of lasting, transposable dispositions which, integrating past experiences, functions at every moment [variable by age, time, and place] as a matrix of perceptions, appreciations and actions" (Garnham and Williams 1980: 213). Each person's habitus is unique, and as you move between fields, your habitus moves with you. In this way your habitus can be seen as an individual structure (comprised of elements of larger social structures) acting upon your agency at a subconscious level. It is at this point that the concept of the sociological imagination becomes crucial. Speaking as president of the British Sociological Association and drawing on Bourdieu, John Holmwood has argued that the point of sociology is to "see structures where others see agency and agency where others see structure" (Holmwood 2013: @21m35s). In other words we can become aware of the structures impacting our choices, and at the same time we have the choice to recognize that those structures have been created by human force and so can be unmade by human forces—including our own. Specifically, if we recognize the field as it currently operates, we can try to deploy whatever capital we possess to try to influence the field. This won't work all the time, perhaps not even part of the time, but recognition of the structures allows us to keep trying.

A key structure in professional fields, according to Bourdieu, is *doxa*, the belief in the game that usually remains unquestioned, and a key aid to recognizing doxa in operation is a *habitus clivé*, a habitus cleft by two worlds, created most commonly due to sudden and extreme social mobility that gives "a unique capacity for reflexivity and self-analysis" (Friedman 2015: 145). From growing up in factories to a transnational professional, I have experienced a particularly sharp change in socioeconomic status. I first realized this when

I moved to the UK as a journalist and discovered I was now relationally "white," whereas previously, in Australia, I had been "ethnic." I wrote this in an email to friends and received a reply from one of my university classmates who said: "Welcome to the white oppressors." However, nowhere have I felt that sense of being a "white oppressor" more than in Goma. This tension in my habitus between past and present was an instance "of misfit . . . that enables the agent to question *doxa*" (Aarseth, Layton, and Nielsen 2016: 148).

The realization that came from the experience of misfit in Goma is why this book is divided into two separate parts. I began this research in the belief that it was our words that needed decolonization—a question of representation. My experience in the borderland of Goma showed me that this was only half the story, that there was an important job to be done in considering decolonization in terms of the social worlds we operate in. Agier has written of the method required in anthropology to understand borderlands as consisting of "the point of view of space, time and relationships . . . of seeing processes and origins, contacts and transformations" (2016: 101). Similarly, chapters 3, 4, and 5 look at space, time, and relationships to consider the processes of foreign correspondence and where transformations might lie for us as professionals, but also for our shared global world.

The distinction between words and worlds, between Parts 1 and 2, is necessary to apply analytic theory to different parts of practice; however, it is also necessary to acknowledge that this distinction is artificial—reality is much messier. For example, in chapter 1 I discuss the propensity of journalists to use a "transactional" charity money-raising frame. This is partly the result of writing conventions inherited from colonial times and partly the result of modern cultural affiliations and shared social geographies working in borderlands with international nongovernmental organizations (INGOs), discussed in chapter 3. Where does one start and the other begin? Ultimately, that question is less important than having the tools to critically engage with both elements of our practice. The various analyses of words and worlds are sutured onto a fairly straightforward chronological display of praxis and knowledge creation. Throughout I ask the reader to follow the journey of praxis and my sometimes missteps down a dawning path of knowledge generation.

Chapter 1 asks the question: How is foreign correspondence from sub-Saharan Africa influenced by writing tropes inherited from the colonial period? Western reporting of Africa has been much discussed and criticized for many years, prominently starting with Hawk (1992). More recent debate has suggested that the negative and colonially inspired reporting frames are overstated, particularly as foreign correspondence has shrunk and more local professionals are hired in Anglophone foreign news production. However, responding to this debate with large-scale content analysis of British newspapers, Chikaire Ezeru (2021) found continued dominance of western bylines and perspectives. Accordingly, the task of decolonizing reporting frameworks remains vitally important. The main data of this chapter is postcolonial literature looking at the writing tropes of the Africanist discourse and postcolonial literature looking at modern media frames. Reading the two together leads to a conceptual map helping to visualize the links between past and present. The lacuna (dark nothingness) trope of the Africanist discourse is identified as most problematic for foreign correspondence. Drawing in analogous analysis from the development field, this leads to my coining of the term the "lacuna effect," the poor reporting that takes place because a "gloomy question mark" (Harden 1991: 14) seems a natural frame of reference for sub-Saharan Africa.

Chapter 2 is a practice-based search for solutions. The problems identified in chapter 1 stemmed from top-down imperialistically tainted knowledge therefore my exploration took an opposite approach and sought to draw on knowledge from refugees themselves. Chapter 2 experiments with a policy framework called "co-design" (Schön and Rein 1994) and seeks to adapt it as a technique for journalism. The Frame Reflection Interview is developed and tested through enlisting former refugees from the east of the DRC now living in Sydney as my "co-designers." By seeking former refugees, not as interviewees, but as interlocutors, the Frame Reflection Interviews serve to break the hierarchy of knowledge created as part of risk society (Beck 2009: 33). The interlocutors broaden the frameworks of understanding that a journalist brings to reporting. This then strengthens the journalist's ability to hear, understand, and represent subaltern voices in a global public sphere. However, this chapter also highlights the struggles encountered trying to introduce new journalism methods to existing structures, including the need to reform my own habitus.

Part 2 shifts the focus from words to worlds. In chapter 3 social geographer David Harvey's concept of a matrix of spatialities is combined with my autoethnographic data in Goma to present novel arguments about the politics of reporting space. Field notes are complimented by literature from international and development studies regarding the existence and impact of Aidlands (Mosse 2011a) and Peacelands (Autesserre 2014). To this interpretation of the space, and the practices it engenders, Bourdieu's theory of doxa is then incorporated to consider how foreign correspondence takes place within the structures of a dominant globalizing class and how this impacts our frameworks of understanding. Of course, an individual correspondent cannot change the powerful and entrenched structures of Aidlands/Peacelands. Instead, the aim is to know these structures better and consider how they impact reportage. This knowledge can then aid the enacting of personal agency toward mitigating negative effects.

Chapter 4 focuses a decolonizing lens on the hero stereotype of foreign correspondence and how this helps to undervalue local journalists and their knowledge. Sparked by ethnographic observation in Goma of the importance of fixers, this chapter picks up on one of the most significant decolonization currents in foreign correspondence, and its main data is the recent critical literature examining the relationship between foreign correspondent and fixer (Armoudian 2016; Murrell 2009, 2015; Palmer and Fontan 2007; Palmer 2018, 2019). This analysis is also structured by the overall context of the growing danger in crisis zones highlighted by many researchers and industry representatives—with that danger growing for both fixers and correspondents. Aided by qualitative research interviews conducted with six fixers in Goma, this chapter analyzes the way in which the neocolonial social relations of Aid/Peacelands intersect with the reporting frames inherited from the colonial library to reinforce the continuance of problematic practices. An argument is made that, contrary to common practice, discussion of foreign correspondent danger in borderlands should, in most cases, be avoided, as this frame can obscure the complex reality faced by locals.

Chapter 5 draws together the knowledge gained via the previous four analyses to consider specific issues in reporting the local-global aspects of borderlands. The importance for understanding the *glocal* nature of social vulnerabilities has already been raised as a key issue in sociology by Beck (2009: 163). Likewise, Agier (2016: 125) states that the local/global dynamic is at the heart of the cosmopolitan condition that should be understood by anthropologists in borderlands. Using specific examples, this chapter argues journalists should also recognize the importance of glocal realities in their reporting. Concepts from previous chapters are applied to analyze how current reporting practices can fall short of glocally focused reporting, and what can be done to move past these blocks. The first example considers the issues of artisanal mining in the DRC, where a lacuna effect is observed in the erasing of the local in favor of the global. In the second example space-time analysis of glocal forces is used to consider issues around journalist-source interactions. Finally, the relationship of foreign correspondents and fixers is re-examined in light of the link between local dangers and global discourses. All three examples point to the crucial role that could be played by foreign correspondence in creating new knowledge around the glocal dynamics of borderlands and how this can empower a move toward grassroots global sensibilities.

In chapter 6 comes the crucial, conceptual discussion of possible reformation of the professional culture and epistemology of global reporting. Using the concept of heterotopia, I argue that gaining a strengthened understanding of the role of foreign correspondence should make it possible to turn the challenges of borderlands into sources of knowledge. This strengthened professionalism requires that journalists, and their publics, consider a new role for international missives; that foreign correspondents become, and come to be seen as, professional cosmopolitans engaged in the creation of a global polity needed for a shared world of risk. The second, and foundational, element needed for this strengthened professionalism is decolonization. This means reaffirming and reframing what is meant by bearing witness. This then leads to Emmanuel Levinas's phenomenological philosophy of the face-to-face relationship with the other. Chapter 6 argues that foreign correspondence can only hope to take up the challenge of reporting borderlands through professionalism philosophically grounded in ethical praxis.

In the short conclusion we return to Fanon and to Levinas. Given the many tasks of decolonization set forth throughout the book, it is necessary to focus on simple actions of what next. And, in fact, it all begins with the face-to-face interview. In a shared world of unequal risks, to see the human without losing sight of the politics, to narrate both at the same time, again and again—this is the path to a shared global polity; these are the storytellers needed right now.

The journey begins.

PART I
WORDS

1

Long, Dark Shadows in Our Heads

Dateline: Athens, Greece, August 12, 2014

James is my savior.[1]

Officially, he is a security guard at the DRC embassy in Athens, but it is he who helps me navigate the forms and personalities in what turns out to be the Herculean task of getting my visa in three days, in time for my prebooked flight.

In embodiment of the belief that persistence beats resistance, I spend the three days camped at the embassy. On that first, sweltering-hot day, in between sorties against the bureaucrats, we sit on the porch out front and talk. He tells me of his life in Kipseli, one of the suburbs of Athens with a thriving migrant population, supposed to be the second most densely populated place in the world. It is a suburb I love, and where some of my best friends, from my mum's little fishing village, live. Kipseli has always been a migrants' suburb: in the age of urbanization it was internal migrants, and now, in the age of globalization, it is external migrants.

James and I discuss relationships, politics, travel, anything and everything, as people with a lot of time and a comfortable porch will.

As the working day draws to a close, James offers me his hand, I look at it, chocolate colored and strong, with that beautiful, familiar patina of lighter points at the joints and at the palm's lifelines, a word flashes through my mind—*Ebola!*

We shake and I feel a split second of fear and internally I recoil. It's ridiculous, I know it's ridiculous, James has not been anywhere near Guinea, Sierra Leone, or Liberia, and at this stage of this outbreak, Ebola has not been reported in the DRC. And yet, as much as I'd like to, I can't pretend to myself I didn't feel that completely unfair, irrational, shameful reaction. As much as I critique, I'm not immune to acts of media-induced irrationality. In fact, perhaps my exposure to so many media reports has made me more susceptible to experiencing them.

Borderland. Chrisanthi Giotis, Oxford University Press. © Oxford University Press 2022.
DOI: 10.1093/oso/9780197565797.003.0002

What did the Ebola coverage tell us about Africa and Africans? Chaotic. Primordial. Primitive. Unreasoning. Incapable. Must be contained. Lives of less value (compare coverage of the few white deaths to those of the wretched of the earth). A continent of darkness and inexplicable dark happenings. FEAR. And then, thank goodness for those white saviors braving such risks to help those poor souls (and keep it far from us). These are the words of Africa living deep in our subconscious and activated by modern media frames whether we want them to be or not.

The Africanist Discourse Is Framed by Foreign Correspondents

The words used by foreign correspondents are not random. Much as, in 1978, Edward Said identified the orientalist discourse to challenge its continuance and validity, various scholars have identified an Africanist discourse (Wainaina 2005; Mbembe 2001; Mamdani 1996; Mudimbe 1994; Madondo 2008; Achebe 1977; Coetzee 2013; Logan 2001; Mafe 2011; Muspratt and Steeves 2012; Brantlinger 1988; Davidson 1992; Miller 1985; Said 1993). This chapter argues that foreign correspondence is still influenced by a particular Africanist discourse created in the west during the colonial period. These Africanist writing frames impact the quality of the words on the page and, I will argue, also impact the specific way in which foreign correspondence takes place.

The main data of this chapter is postcolonial literature looking at the writing tropes of the Africanist discourse and postcolonial literature looking at modern media frames. Africanist tropes are writing metaphors that were commonly used during the colonial period in both fiction and nonfiction works. The similarities and differences with journalism story frames will be discussed further on. Reading the two together leads to a conceptual map helping to us visualize the links between past and present. The lacuna (dark nothingness) trope of the Africanist discourse is identified as most problematic for today's foreign correspondence, as it normalizes a lack of comprehension of the dynamics of the borderland in the east of the Democratic Republic of Congo (DRC) and of African borderlands in general.

The purpose of focusing on the continuance of the Africanist discourse is to remove the problematic frames that obscure knowledge and limit our lexicon. Accordingly, we need to identify the key mechanisms of how the

Africanist discourse continues to surface in modern journalism. My focus is on the Africanist discourse because my case-study journalism took place in the DRC; however, the principles and method underpinning this decolonization process can be applied to postcolonial borderlands everywhere.

In focusing on problematic aspects of media frames in need of decolonization, I will be rightly accused of negative cherry-picking. As noted by Scott, this creates a danger:

> Repeatedly emphasising only the anticipated and problematic aspects of representations of Africa may, inadvertently, end up serving to reinforce the very same ideas that these studies often seek to challenge. If nothing else, generalised critiques about the apparent limitations of all news coverage of Africa can inhibit constructive dialogue with those responsible for producing such coverage. (2016: 40)

Taking into account Scott's critique, I wish to particularly point out that this is not a content analysis of current journalism practice; nor does it seek to prove that the frames related to Africanist discourse are the only frames through which Africa is understood. There is plenty of Anglo-American foreign correspondence that isn't structured in Africanist discourse. But this does not absolve us of decolonization work for the profession. I also hope that in later chapters, by tying this decolonization work to my own practice, a productive conversation becomes possible. Said once said of his work: "Nowhere do I argue that Orientalism is evil, or sloppy or uniformly the same in the work of each and every Orientalist. But I do say the guild of Orientalists has a specific history and complicity with imperial power which it would be Panglossian to call irrelevant" (1995: 342). If I can appropriate Said's words, nowhere do I say all foreign correspondence of sub-Saharan Africa is evil, sloppy, or uniformly the same, but as foreign correspondents, we have histories of complicity, first with imperial power, then with Cold War realpolitik, and then with the INGO industry, and this complicity is not irrelevant. In fact it is perhaps more relevant than ever.

Recent scholarship has highlighted the entanglement of journalists and journalistic practices with international nongovernment organizations (Franks 2010; Powers 2015, 2016; Rothmyer 2011; Wright 2018), and this chapter finishes by considering how the story frames used by journalists interact with problematic assumptions in humanitarianism, reinforcing the colonial legacies of both professions. A new concept is introduced, the "lacuna effect" in journalism, which draws upon and compliments the "white man's burden effect" already identified in humanitarianism. Much as the white man's burden effect places blame for development failures on aid recipients, in the process blinding interveners to their own faults, the lacuna effect is revealed when journalists consider it normal to write stories with very little context and background information because a "gloomy question mark" (Harden 1991: 14) seems a natural frame of reference for sub-Saharan Africa. Moreover, the lacuna effect blinds journalists to their ignorance, so the faults are repeated.

The work of decolonization lies in unpacking our profession's sociocultural trajectory of complicity. The first point to understand is that Africanist discourse did not develop in a representational bubble—quite the contrary. Scholars have shown that the tropes took on particular forms in line with the western needs of exploitation and colonization. For example, the burgeoning rubber trade was fed by forced labor. However, enlightenment ideals meant no man could enslave another, so "the order is given to reduce the inhabitants of the annexed country to the level of superior monkeys in order to justify the settler's treatment of them as beasts of burden" (Sartre 1963: 15), and journalism played its part. The DRC is unique in the colonization story of Africa, as it was colonized as the property of the Belgian King, not as the property of the Belgian nation. Furthermore, it was "opened up" for colonization by a journalist—in other words, a person very aware of the power of words and of the media. Henry Morton Stanley's trip, funded by King Leopold, was a sort of reality show of the time, and so aware was Stanley of the importance of the representations he would frame, that he required all of his white companions traveling with him to sign contracts promising they would not publish their own accounts of the trip until "well after" Stanley had published his version of events (Dunn 2003: 24). By doing this Stanley gave himself the power of the "primary definer" (Hall 1978) of events in the media, against which all subsequent versions of events must contend. As part of his production, Stanley highlighted the idea of the Congolese people as animals. He often used animal metaphors and "referred to Africans as 'beasts' and 'apes' and compared them to dogs, often as the canine's inferiors" (Dunn 2003: 29).

Stanley was assisted in this task of making the animal connection both by social Darwinism and by preexisting Africanist discourse, which was shifting in line with the shifting needs of colonization:

The relatively naïve racism of the early decades of the [nineteenth] century often found room for the noble savage, as in several of Marryat's novels; increasingly it gave way to depictions of Africans and Asians in terms of pseudo-scientific racism similar to Robert Knox's, based on reductive versions of social Darwinism. (Brantlinger 1988: 39)

Mahmood Mamdani makes it clear that this loss of nobility in fictional pages corresponded with a real-life reversal in the standing of the African as a citizen of modernity: "Although Africans held nearly half of all senior official posts in Sierra Leone in 1892, their share declined to one in six by 1912. The door that was shut in the face of the educated strata from the start of the scramble remained closed until the final chapter of colonial rule" (Mamdani 1996: 76). This research does not include a detailed economic analysis of the colonial era, but that doesn't mean that the logic of exploitative economics wasn't there, playing its part. It is important to constantly bear in mind that representation practices cannot be considered in isolation from the practices of colonialism.

The second point that needs to be constantly borne in mind is that "discourse" can be understood as a family of related statements, and just as family members tend to support each other, so too the different tropes of the Africanist discourse, connect, support, and reinforce each other—unfortunately. This is true, even when they seem contradictory, because they still belong to the same family of statements. For example, Nothias (2014) found that the positively framed "Africa Rising" news coverage still contained colonial tropes. Hall explains:

Since in order to say something meaningful, we have to "enter language," where all sorts of older meanings which pre-date us, are already stored from previous eras, we can never cleanse language completely, screening out all the other, hidden meanings which may modify or distort what we want to say. There is a constant *sliding of meaning* in all interpretation, a margin— something in excess of what we want to say—in which other meanings overshadow the statement or the text. (Hall 2013b: 17–18)

Within the many writings of scholars concerned with the way Africa and Africans are represented, tropes, or metaphoric ways of writing, are identified as crucial for structuring this Africanist discourse, and for providing the colonial echoes that continue to influence the western misunderstanding of African issues. In the postcolonial literature there is also a clear link between the tropes of novels written during the heyday of colonization and the journalism of today. However, in journalism we do not often consider writing tropes, which is a term associated more with cultural studies. Instead we are more likely to consider stories as being "framed." And frames too contain metaphors. Robert Entman argues that "news frames are constructed from and embodied in the keywords, metaphors, concepts, symbols, and visual images emphasized in a news narrative" (1991: 7) and further highlights the power dynamic involved in that construction:

> To frame is to *select some aspects of a perceived reality and make them more salient in a communicating text, in such a way as to promote a particular problem definition, causal interpretation, moral evaluation, and/or treatment recommendation* for the item described. (Entman 1993: 52)

These framing effects carry responsibilities—particularly when reporting from places where audiences are likely to have little first-hand experience for themselves:

> Journalists are often doing more than describing and "reporting," when they are reporting from dangerous places; through their crafted prose, dramatic visuals and often affective storytelling, they are in fact, through their journalistic practice, inviting readers to engage, empathetically, as well as cognitively and analytically. (Cottle, Sambrook, and Mosdell 2016: 122)

It is this responsibility that means each word used must be understood in its sociopolitical context. It is easy to forget that most correspondents, when they first arrive, have little experience in postcolonial contexts—and it is here where there is danger that they will draw on narratives and interpretations that are part of Africanist discourse—if for no other reason than these are the cultural products that they have had access to up to that point. In discussing the growing trend for "melodramatic news narratives that feature the war correspondents as heroes" Palmer highlights:

> It is important to remember that these narratives are larger than any one individual. They draw upon diffuse cultural traditions that are deeply entrenched in the news industry, as well as in western, Anglophone society.

This is why it is essential for critical media scholars to zoom out from the news narratives themselves and examine them in the broader context of production culture—a culture that cannot exist outside the larger sociopolitical culture in which it operates. (2018: 22)

Palmer's focus is the war on terror and the Middle East, and the distortions in coverage created by Anglocentrism. My focus is on sub-Saharan Africa and on the distortions created by Africanist discourse. The clear overlap is in an analysis that sees the logics and practices associated with foreign correspondents and war correspondents as influenced by the larger sociopolitical culture in which the journalists, and their news organizations, exist. Africanist discourse is part of that sociopolitical context.

The way in which Africanist writing tropes overlap and differ from modern media frames is crucial to enhancing our understanding of the structures at work, and will be unpacked in detail further on in this chapter. Detailed discussion is important for embedding our understanding; however, there is also the question what simple tools we can create so as to identify that overarching structure and enact our agency to try to shift it. A variety of postcolonial scholars, from different disciplines, have done work identifying Africanist discourse in modern media frames. It seemed to me it would be helpful to use that literature to develop an easy-to-reference typology of tropes from the colonial period and their equivalent modern-day frames (Table 1.1).

Table 1.1 is tool 1. Identification is an important aspect of decolonization. However, there is also the question of how and why these frames get used; thus a further tool is needed. To better understand this second aspect of the question, I undertook a small qualitative frame analysis. This analysis was of all African stories by the Australian Broadcasting Corporation's (ABC) flagship TV program *Foreign Correspondent* in the years 2008 and 2013 and all stories of the DRC throughout 2008–2013.[2] The point of the qualitative analysis was not to add to the typology identified in the table but to understand the dynamics of how frames come to life in stories and how they evoke the Africanist discourse.

Each story was analyzed using three different theories of framing. I searched for how frames can be created by selection and repetition, that is, identifying different themes in the story that the journalist selected for attention and then counting seconds of reportage per theme to determine which elements were given prominence via repetition. Second, I analyzed the

Table 1.1 Africanist Discourse Tropes and Frames

Trope	Media frame
Lacuna	Backdrop
Animal	"Animal Eden"
Animal	"Save the animals"
Evil, tribal (irrational) sexual savage	*Heart of Darkness*
Evil, tribal (irrational) sexual savage	Tribalism
Evil, tribal (irrational) sexual savage	Chaotic-tribal/unreasoned sexual violence
Evil, tribal (irrational) sexual savage	Big man (+ *development promise*)
Evil, tribal (irrational) sexual savage	Victim/poverty porn
White hero/white man's burden	White hero and/or angel of mercy
White Hero/white man's burden	Foreign correspondent's dangerous journey (*also variation on backdrop frame*)
White hero/white man's burden	Transactional (+ *development promise*)
White hero/white man's burden	Uber-westerner and the uber-victim (+ *development promise*)

programs looking to see which cause(s) and solution(s) were offered to the audience. Third, I searched for overarching metaphors. I chose these three techniques as they are commonly referred to in journalism studies literature on how frames are created and serve as working routines for journalists (Entman 1991; Iyengar 1994; Scheufele 2000). I also looked at production issues that may have affected the choice of the frame used. For example, was there an exclusive interview or were there specific time constraints on the making of the story?

I chose to focus on TV reportage, even though the colonial library is based on literature and my own medium of choice for journalism is print. The issue of medium and genre does not impact the discovery of tropes, as these metaphors of storytelling easily transfer across genres and mediums. Furthermore, it is important to look at what the industry leader was doing, as this then influences other journalists. The ABC is Australia's national broadcaster, our equivalent of the BBC, and its foreign correspondent network was modeled on the BBC. Although its foreign correspondent network is much diminished in recent times, it is still considered an industry leader. The ABC is consistently Australia's most trusted news brand (Roy Morgan 2019). Its TV news channel is bucking the western media trend of declining TV viewership, with increasing audience numbers, mirroring the audience growth of its online offerings, and two of the three most downloaded video-on-demand current affairs programs in 2021 were episodes of *Foreign Correspondent* (ABC 2021).

All episodes analyzed in my sample contained more than one example of Africanist discourse, and it was this realization that made me go back to the literature, this time looking not for tropes and their equivalent frames but instead focusing on the relationships between all the tropes and frames so as to excavate a potential driving dynamic. I identified the lacuna trope as doing the most work in terms of keeping the Africanist discourse alive in modern journalism.

This second analysis of the literature resulted in Figure 1.1. Note that the media frames of Africanist discourse are not only reinforced by the tropes but, in turn, connect back to the tropes, reinforcing those tropes in a vicious circle of continuity. Also note the overlap between the two blocks of media frames; they may be in a different grouping, but they reinforce each other's continuance through existing together in media discourse of sub-Saharan Africa.

The diagram summarizes which tropes relate to which media frames and how they connect to and reinforce each other—with some tropes and frames more connected than others. I hope it will help readers keep in mind the

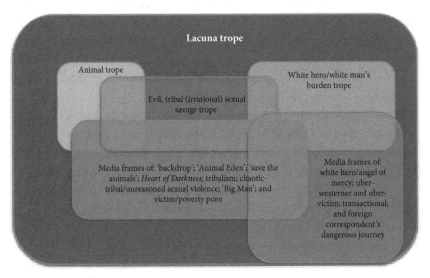

Figure 1.1 Africanist Discourse Tropes and Frames

overlaps of the discourse. The diagram also highlights the role of the lacuna trope in providing a strong narrative bed for other tropes to rest upon.

The Africanist discourse literature points to the obscuring of knowledge effects for Africa in general—not borderlands—but it is clear to see that problematic reporting frames used in discussing war and politics are highly relevant for our study. Accordingly, I will not go through every one of the linked tropes and frames.[3] Instead, I will focus on the tropes and frames that are most relevant for reporting in borderlands. These colonial ghosts must be excised to better understand, and report, the postcolonial present of unending wars.

My discussion takes discourse as a starting point and not an end. After looking into the detail of the key tropes and frames, I will then argue that the lacuna trope has had a lacuna effect on foreign correspondence. The discussion will make clear that Africanist discourse impacts practice beyond the words on the page.

But first, words.

The Lacuna Trope

In Africanist discourse the style of writing that I have classified as belonging to this lacuna trope can best be understood as writing that uses "dark nothingness" as a metaphor for Africa, the type of metaphor that created the once-common description of Africa as the "dark continent." I have adopted the term "lacuna" from Achille Mbembe, who writes: "It is not true, as either starting point or conclusion, that Africa is an incomparable monster, a silent shadow and mute place of darkness, amounting to no more than a lacuna" (2001: 9).

This dark nothingness consists, first, of a dark mystery, a void in the knowledge of the westerner reading about Africa, and it is in relation to this mysteriousness that the lacuna becomes particularly *useful* to western writers, such as foreign correspondents, who wish to create dramatic narratives. This lack of knowledge creates a useful ambivalence. Both Miller (1985) and Mudimbe (1994) emphasize that there were many different conceptions of Africans in ancient Greece. In Homeric verse the Ethiopians, then a blanket term for much of Africa, were viewed as the favored of the gods and the "justest men whom the Gods leave their abode frequently to visit" (Miller 1985: 24). But where Homer eulogized, Herodotus reported "dog eared men and the headless who have eyes in their chests" (Miller 1985: 3), a description repeated by a "London merchant called John Locke, who sailed to west Africa in 1561. . . . After referring to the black Africans as 'beasts who have no houses,' he writes, 'They are also people without heads, having their mouth and eyes in their breasts'" (Adiche 2009: @06m34s). Because Africa was the "dark continent," you could say whatever you wished about it.

This role of black persons as others about whom you could say whatever you wanted, and who thus had no existence of their own, is given voice by Fanon in his 1952 work *Black Skin, White Masks*: "The black man has *no ontological resistance* in the eyes of the white man" (1986 [1952]: 110; my italics). I have highlighted ontological resistance because there is a difference between "Othering" as understood through orientalist discourse, and the lacuna project of Africanist discourse, and this difference is important. It is a theme that Christopher Miller develops in *Blank Darkness*, his book on the way Africa is presented in French writings. Miller states: "The etymologies [of the word "Africa"] of 'colony' and 'Afer' tell exactly the same story of an empty slate, written on by outsiders" (1985: 16).

Miller says that this nullity, which is embraced by Africanist writers, allows them to turn Africans into objects of "soft wax" and fashion them in any way they see fit. He gives two striking examples from France, where Africans, arriving in Europe, were first praised and feted and then vilified as the mood of the populace changed. He then quotes Douglas Grant in describing a British example of the same phenomenon: the person "existed only so far as he had been interpreted according to current notions of Africa" (1985: 39).

All this of course means that we are not studying Africa itself, but, as V. Y. Mudimbe puts it, *The Idea of Africa*, its construction through western texts and its "culmination in the colonial library" (1994: 213). That colonial library, especially in the early days, emphasized adventure stories: "The adventure tales that formed the light reading of Englishmen for two hundred years and more after *Robinson Crusoe* were, in fact, the energizing myth of English imperialism" (Green in Brantlinger 1988: 11; see also Said 1993: xi, 12). As such, they represented an Africa in relation to the imperial project rather than presenting writing on the people and places of Africa itself.

Even when not supporting colonial practices, the lacuna trope remains. Joseph Conrad's best-known work, *Heart of Darkness*, is generally considered a critique of colonial practices and has been described as "very great" with a deserved "canonical place in the legend of modern literature" (Trilling in Brantlinger 1988: 269). However, the novella does nothing to recognize the populace of the Congo as human beings with their own valuable alterity and a right to self-determination on their own terms (Said 1993: 22–25).

In an influential critique, Chinua Achebe drew attention to the role of the dramatic, dark mysterious lacuna trope as a perfect backdrop to the novella and the "preposterous and perverse kind of arrogance in thus reducing Africa to the role of props for the breakup of one petty European mind" (1977: 788). This casting of Africa as a backdrop had real consequences for the type of imperialism practiced. Fanon writes: "Under the German occupation the French remained human beings. Under the French occupation the Germans remained human beings. In Algeria there is not simply domination but the decision, literally to occupy nothing but territory. The Algerians, the women dressed in haiks, the palm groves, and the camels form a landscape, the *natural* backdrop for the French presence" (Fanon 2004 [1963]: 182).[4]

Two techniques through which the lacuna trope was written and maintained included denying African people their voice and history. Achebe criticizes Conrad's lack of African speaking characters, and Sven Lindqvist in *Exterminate All the Brutes* describes this lack of voice as the norm:

> What did the king of Benin feel as he was hunted like a wild animal in the forests while his capital was going up in flames? What did the king of Ashanti feel as he crawled up to kiss the boots of his British overlords?
>
> No one asked them. No one listened to those whom the weapons of the gods subjugated. Only very rarely do we hear them speak. (1998: 61)

Eric Wolf in his acclaimed work *Europe and the People without History* points out that this denial of voice is reinforced through the power to bestow and deny history. Wolf says that "a history written by slavers and their beneficiaries has long obliterated the African past . . . [denying] both a complex political economy before the advent of the Europeans and the organizational ability exhibited by the Africans in pursuit of the [slave] trade once begun" (1982: 229). This denial of history, which sits inside the nothingness of the lacuna, in turn creates what Mbembe calls *facticity*:

> By *facticity* is meant that, in Hegel's words, "the thing *is*; and it *is* merely because it *is* . . . and this simple immediacy constitutes its *truth*." In such case, there is nothing to justify; since things and institutions have always been there, there is no need to seek any other ground for them than the *fact of their being there*." (Mbembe 2001: 3–4)

The *facticity* element of the lacuna trope is particularly important for foreign correspondents to be wary of as it connects to the tribalism trope to be discussed presently.

Media Frame of "Backdrop"

To see the continued subconscious life of the "blank canvas" trope and its difference from the "Othering" trope applied in Orientalism, you need look no further than the advertising frames applied to tourism. Think of Egypt, Jordan, India, China, and Japan, and the way these places are presented in tourism media as alternatives to cultural tourism in Europe. Then think of the "adventure tourism" frame applied to Africa, a frame that once more presents Africa and Africans as an exotic backdrop against which white people can live out their paid-for heroic fantasies. Journalists use almost exactly the same background frame that tourism brochures apply when, as part of their stories, foreign correspondents emphasize the "adventurous" terrains through which they travel.

In my frame analysis of African stories in the ABC TV flagship program *Foreign Correspondent*, in the two years of 2008 and 2013 I found that in *every* episode involving sub-Saharan Africa (five in 2008 and three in 2013) there was at least some element of repeated emphasis placed on the dangerous/difficult journey undertaken by the correspondent. In one 2013 episode the backdrop frame was present in almost 15 minutes of the 28-minute program, with the correspondent repeatedly emphasizing he was "discovering Mali on a difficult journey to Timbuktu." This was in a program ostensibly about the recent Islamist-led coup in Mali and the potential for its revival after the withdrawal of French troops. However, the journey to Timbuktu has become such a staple of foreign correspondence that is seems hard for journalists to talk about that region and *not* talk about the journey. This tendency is noted by Hannerz, who writes of "Dateline Timbuktu," noting "a remarkable number of correspondents, too, sooner or later make it to Timbuktu" (2004: 131).

The "backdrop" media frame is easy to identify when the foreign correspondent is focusing on the difficult, adventurous journey. However, it can also be used when the journalist is stationary, and it seems to be particularly prevalent in reportage from refugee camps. Both cultural critic Wainaina (2005) and foreign correspondent Keane argue African refugees are cast as backdrops. Keane describes the situation thus:

Since the end of colonialism, Western correspondents have stood in front of emaciated Africans or piles of African bodies and used the language of the Old Testament to mediate the horrors to their audiences. . . . Viewers at home are watching (usually) a white reporter and white aid worker, and beyond them almost as backdrops are the wretched African masses. . . .

I have many personal memories of such scenes. Among my least favorite was the sight of the American vice president's wife, Tipper Gore, descending on Goma as an army of press paraded her compassion in front of a backdrop of Rwandan Hutu refugees. (2004: 9)

Muspratt and Steeves call this practice "erasure journalism" (2012)—however, they highlight that Africans are not just erased, they are also *replaced* by white people like Tipper Gore—an important point to bear in mind when looking at the white hero / white man's burden trope and its related frames later on.

This practice of casting Africa as a background to play out the west's "heroics" was given a different spin during the Cold War; at this time the media metaphor for Africa was that of a chessboard on which the United States and the USSR played their game. According to respected journalist and historian of Africa Basil Davidson, this distorted reality: "African movements that seldom or never heard more of Communism than the names of Marx and Lenin, or even less, were solemnly denounced [by western commentators] as potent agents of the red revolution" (1992: 224).

After the Cold War the chessboard-background narrative continued, albeit under the new guise of the "clash of civilizations." In describing research on Darfur that he was involved in for the African Union as part of the Abuja ceasefire negotiations, Mamdani notes:

The violence had spread over two axes: a north-south axis that pitted nomadic against peasant tribes, and an east-west axis, in the south, that led to two kinds of nomadic tribes—those with homelands and those without—confronting one another.

But [the campaign] Save Darfur, and the media in thrall, focused exclusively on the north-south axis, thus creating a further distortion—that this was a conflict between "Arab" and "African" tribes. (2009: 36)

Gruley and Duvall highlight the same "clash of civilizations" misrepresentation in their frame analysis of Darfur reportage from the *New York Times* and *Washington Post*. They conclude:

The historical and political factors that have only recently caused "Arab" and "African" identities to become salient in Darfur were mostly overlooked in NYT and WP coverage. . . . Indeed, the political motivations for the rebellion were frequently neglected or simplified. (2012: 37)

Of course, the capitalism-versus-communism narrative of the Cold War and the more recent clash-of-civilizations narrative aren't just limited to Africa. Williams (2019: 20) notes that "indigenous roots [of wars] were regularly ignored" by the Cold War frame applied "across continents." He goes on to discuss that warfare after the fall of the Berlin Wall only become more difficult for western war correspondents to understand and report, which led to attempts to restore order through a new binary: the war on terror. The fact that these forces of framing also have sway in other postcolonial states does not nullify the argument of the lacuna trope; it simply means that when these frames overlap with sub-Saharan Africa, they help to keep Africanist discourse, and its underlying assumptions, alive.

Heart of Darkness: Trope to Frame

Reading Fanon, I find a relationship between the lacuna, or "negation," of the colonized and the insistence on darkness/evil:

> The colonist is not content with stating that the colonized world has lost its values or worse *never possessed any*. The "native" is declared impervious to ethics, representing not only the absence of values but also the *negation* of values. He is, dare we say it, the enemy of values. In other words, *absolute evil*. (Fanon 2004 [1963]: 6; my italics)

So what was once blankness became a different sort of unknown, the unknown of deep, dark evil. And it is here that we find the lacuna reinforcing itself in a new way, for it is not for the civilized European mind to plumb the depths of evil. A shroud of mystery descends on the continent, and its alleged incomprehensibility becomes an accepted part of the way it is viewed.

The metatext of this trope of writing is of course *Heart of Darkness*, partly because it is one of the few stories of Africa that is widely read in the west. This explains why the DRC is uniquely "over-textualised . . . a symbolic stand-in for sub-Saharan Africa" (Dunn 2003: 8). This also helps explain why references from *Heart of Darkness* continue to find their way *directly* into modern journalism. In his influential critique of the book, Chinua Achebe noted Conrad was "pretending to record scenes . . . [but] in reality engaged in . . . a bombardment of emotive words and other forms of trickery" (1977: 784).

Achebe notes: "The eagle-eyed English critic, F. R. Leavis, drew attention nearly thirty years ago to Conrad's 'adjectival insistence upon inexpressible and incomprehensible mystery.'" This, according to Achebe, was possible because "Conrad chose his subject well—one which was guaranteed not to put him in conflict with the psychological predisposition of his readers or raise the need for him to contend with their resistance. He chose the role of purveyor of comforting myths" (1977: 784). In other words Conrad chose to continue the Africanist discourse even while he was attempting to dismantle the system of colonization from which it emerged.

The *Heart of Darkness* frame is used in a similar way in journalism and is most likely to occur when the journalist has little experience of the continent.

Describing the first Ebola outbreak to make international headlines, in Zaire in 1995, the *New York Times* correspondent Howard W. French writes: "Reporters had come from everywhere, *many of them getting their first taste of Africa* in what was possibly the most dilapidated and confusing country on the continent. In the process more than a few of them were overworking clichés drawn from *Heart of Darkness*" (2005: 49–50; my italics). BBC correspondent Fergal Keane similarly writes, "The cliché most favoured by *distant headline writers* who coin phrases about Africa . . . [is] 'the Heart of Darkness'" (Keane 2004: 8; my italics). The important point I wish to highlight here, which both correspondents reference, is that the *Heart of Darkness* frame is a reporting tool used most often by those with the least knowledge and this feeds into the structural issues in the profession, including parachute journalism and the fact that the final product is often framed by those higher up the editorial chain who may have no knowledge of African contexts. Nothias interviewed 35 foreign correspondents in Kenya and South Africa and found that attempts to present more nuanced descriptions of Africa that avoid stereotypes were often undone by headline writers and editors in home offices (Nothias 2020: 253–55).

Tribalism, Chaos, and Sexual Violence

From an unknowable dark land to one populated by tribal, irrational, sexual savages, the links in the narrative write themselves, and modern media frames bear the traces.

J. Gruley and C. S. Duvall, academic geographers, set out to "determine if tribalism remains important in coverage of African content" through a framing study of the Darfur conflict as represented in the *New York Times* and the *Washington Post* between 2003 and 2009. This was an important investigation as "Western news media have represented conflicts in Africa as 'tribal,' a trope that erases geographic and historical context, and discourages actions that could prevent or reduce violent conflict" (Gruley and Duvall 2012: 29). In citing the erasure of geographic and historical context, the authors helpfully point to the link between the lacuna trope and the tribal trope, and further quote various postcolonial scholars to conclude that "the imagined, tribal Africa is timeless and unchanging, a place without history" (2012: 30).

These frames have implications for real-world policy. Discussing Liberian war coverage, Atkinson writes:

> Media concentration on primordial ethnic identity as a cause of war, with its apparent manifestations in savagery and even cannibalism, helps to obscure critical political or economic factors driving the violence. It contributes to increasingly popular misconceptions of African wars as being fought for primitive causes beyond the influence of the west. (1999: 192)

Seaton speculates that simplistic ethnic and tribal interpretation of contemporary warfare has contributed to the poor understanding, and reportage, of the role of modern peacekeeping missions:

> In highly involved situations that they have little time to understand . . . some journalists have used "ethnicity," tribalism" or "history" to explain contemporary conflicts . . . the politics of reporting peace-keeping missions . . . is as undeveloped as the interventions themselves seem to be. (1999: 59)

In the Darfur study, Gruley and Duvall found that despite some counternarratives, especially as the conflict drew on, an overwhelming number of articles framed the stories in terms of tribalism. This reinforces "facticity." Mamdani, in writing about the response of New Yorkers to the Sudan conflict, and contrasting it with their response to Iraq, says:

> Iraq is a messy place in the American imagination, a place with messy politics. Americans worry about what their government should do in Iraq. Should it withdraw? What would happen if it did? In contrast, there is nothing messy about Darfur. It is a place without history and without politics. (Mamdani 2007: online)

Ironically, this sense of simplicity—driven by facticity and linked to the dark-continent trope—is counteracted by the overt message of that trope: that political irrationality is a natural part of the African makeup and thus conflicts are too confusing to understand. This is not lost on African political leaders, who use mimicry of colonial tropes for their own ends. Gruley and Duvall write:

> Khartoum has repeatedly called the conflict "entirely tribal," in order to deny its involvement and discourage third-party intervention.... Tribalism destroys nearly all aspects of geographic reality, enabling either unreason, or a *chaos* of reasons as varied as those doing the representing, to become the imagined, and unassailable, context of violence. (2012: 42; my italics)

I italicized "chaos" because this frame has been highlighted as particularly problematic by respected longtime Africa foreign correspondent, and now Royal African Society director, Richard Dowden. In his 2009 book *Africa: Altered States, Ordinary Miracles* Dowden writes:

> Africa may look like chaos and madness but there is always a comprehensible—if complex—explanation. A group of us, journalists who covered Africa full time, decided that we would ban the word chaos from our reporting.... Our watchword was, "If you describe it as chaos you haven't worked hard enough." (2009: 5)

Certainly, many Kenyans believed western journalists weren't working hard enough when they flew in to cover the 2013 Kenyan election, or if they were working hard, it was to reaffirm the existing tribal/unreasoned/chaotic violence frame. The previous Kenyan election, in December 2007, led to a disputed result and two months of violence that left around 1,000 people dead.

It was perhaps natural for reporters to put the 2013 election in this context and wonder whether there would be a repetition of the violence. However, five years had passed, during which time a unity coalition government had been in power and the political climate had changed. Whether there were mitigating factors or not, Kenyans felt they were being unfairly framed by foreign journalists, and they made their views known on Twitter with the critical (and humorous) hashtags #SomeoneTellCNN, #PicturesForStuart, and #TweetLikeAForeignJournalist. The first hashtag had been used previously on Twitter to critique CNN reporting in Kenya and was revived after CNN seemed to be sensationalizing the possibility of tribal violence. The second, #PicturesForStuart, was also in response to the sensationalization of supposedly "chaotic" voting scenes (Dewey 2013).

This response on social media may seem disproportionate to the crime. Reporting the possibility of election violence is certainly reasonable considering the events that followed the 2007 elections. However, it is worth noting that the 2013 social media attack on western journalists focused on the *framing*, not the facts reported. It was the framing of 2007 that had stayed with Kenyans, framing described by one academic as " 'atavistic' tribalism [that] carried echoes of Stanley and other early Western visitors" (Rothmyer 2011: 18). The Kenyan Twitterverse (#KOT—Kenyans On Twitter) gave foreign journalists their verdict on the 2007 framing.

When Gruley and Duvall began their study of Darfur, they were influenced by criticisms of journalism raised after the coverage of the Rwanda genocide. That event was reported as an archetypal tribal war, and western journalists paid scant attention to other factors, including politics (Gruley and Duvall 2012: 30). This tribal frame also ignored the fact that before the period of Belgian colonization Hutu and Tutsi would be better described as different economic classes rather than tribes, with "membership in [the] elite [the] basic determinant of 'Tutsi' identity" (Dunn 2003: 148). It was under Belgian colonial rule that the "division between the two groups became increasingly reified" (Dunn 2003: 148), as happened in other areas of Africa.

The continuing conflict in the east of the DRC has been heavily intertwined with the Rwandan genocide. In a demonstration of how closely linked the tribalism and chaos frames are, Dunn writes:

> Throughout the 1990s, media coverage [of the DRC] was steeped in the language of "New Barbarism"—employing the rhetoric of "chaos," tribalism, and irrational African violence. . . . Zaïre was represented as wallowing

in the "law of the jungle" (*New York Times*, 17 March 1997) . . . *Newsweek* greeted the collapse of Mobutu's regime with a headline lifted directly from Conrad's *Heart of Darkness*: "The Horror, The Horror" (31 March 1997). (2003: 166)

Atkinson links the new barbarism thesis to real issues in borderlands, including the overlooking of illegal resource mining as a driving force of war. Atkinson states that in Liberia "effective action to resolve the Liberian conflict was successful only following serious attention by the international community to the issue of the illegal economy" (1999: 214). Outside Africa, tribalism is known as ethnicity and became a common explanation for wars around the world in the post-Cold War 1990s. This framing is "at best misinformed and misleading" (Allen 1999: 33). Allen highlights that this leaves many unasked questions about the nature of contemporary warfare. Most obviously—for journalists—what role did the collapse of state institutions and the proliferation of arms play in the new wars of the 1990s, and—for journalists and media scholars—what role does the media play in reinforcing, or even recreating, ethnic tensions?

Moving from the tribal, chaotic and unreasoned violence (over)reported in foreign news frames to the sexual violence frame, we again find the DRC is *the* place to look. In 2010, Margot Wallstrom, the UN's special representative on sexual violence in conflict, called the DRC the "rape capital of the world." It was a media-savvy line that drew attention to the conflict, but in fact this issue of rape in the DRC had been helping create headlines for several years, with a

singular focus on sexual violence [in] reports, articles, news clips, appeals and documentaries dealing specifically with the issue of rape. Other forms of violence—mass killings, systematic torture, forced recruitment, forced labour and property violations, etc.—are committed on a massive scale but receive far less attention and resources. (ErikssonBaaz and Stern 2013: 6)

The creation of specific narratives around particular kinds of horrors within violent crises has become a common occurrence in sub-Saharan Africa. These are based on the real effects of the wars. In northern Uganda, where abduction was a real issue, media reports repeatedly centered on the nightly migration of village children to sleep in the safety of city streets (Allen and Vlassenroot 2010: 3). In the conflict in Liberia the media spotlight fell on the use of child soldiers, and in the conflict in Sierra Leone it was on the high number of amputees. So, in some respects, the singling out of rape victims as the "emergency within the emergency" is ordinary NGO and media practice (Eriksson Baaz and Stern 2013: 96).

The fact that a particular focus on specific horrors is normal INGO and media practice does not negate the interplay between the "rape capital of the world" frame and the Africanist sexual savage trope. Former *New York Times* correspondent for West Africa and Columbia journalism associate professor Howard W. French posits that the rape frame "allows us, meaning the general public, to become interested in Africa in ways that respond to some pre-existing notions that we have of Africa" (French 2011: @01m12s). Relatedly, feminist development scholars Maria Eriksson Baaz and Maria Stern write: "The acts of rape that have occurred in the DRC are often represented as a result of the supposed bestiality of the rapists. While rarely directly stated, this is intimated through not so subtle allusions" (2013: 25).

Big-Man Frame

At the same time they strengthened or indeed created "tribes," colonial powers strengthened and ordained chiefs to exercise the colonialists' indirect rule. The transition from describing all-powerful chiefs in Africanist discourse to describing all-powerful political leaders in media frames happened seamlessly. This transition is seen, for example, in a book designed to help journalists write about postcolonial Africa. The introduction of the *Reuters Guide to the New Africans*, written in 1967, is subtitled "Men Who Transformed Africa," and its opening lines state: "The story of Africa's sudden surge to nationhood—200 million people have won independence within the last decade—is told by the lives of Africa's leaders" (Taylor 1967: 7).

Focusing on the lives of leaders reinforces the notion of a "chief," so it is easy to see how the frame of the big man came to be such a regular feature of African reporting. Of note here is that the term "big man" was once employed by foreign correspondents in many different places around the world, but in Africa, perhaps assisted by the continuing influence of the Africanist discourse, the term has stuck:

Quite recently, Latin America had been full of Big Men, as had Eastern Europe, and much of Asia for that matter, but only in Africa did the term—actually borrowed from anthropologists' descriptions of Pacific Island societies—become a fixed moniker employed by writers too bored or lazy to get beyond such labels. (French 2005: 93)

Too bored or lazy? Or perhaps too unaware of the dark shadows from the colonial library telling them that big man / chief is the "natural" frame to employ in a story about apolitical/tribal Africa.

In the big-man frame, we again see the DRC as exemplar. Dunn argues that the United States' Cold War–motivated support for President Mobuto during the Zaire era was based on the belief that chaos was the "fertile soil from which 'red weeds' grow" (2003: 86), and as chaos was the "natural" condition of any African country without a "chief," a strong leader / big man like Mobuto was needed. Dunn further argues Mobuto's strategy of *authenticité*, typified by his leopard skin hat and carved cane, can be seen as a strategy of mimicry in which he used the colonial trope of "Congo = chaos" in his own favor by extending the equation to add "Mobutu = stability" (2003: 113–22).

A variation on the Mobuto = stability theme was Mobuto = development. This version, again tied to the tribal/apolitical trope of Africa, relied on an assumption of linear development that placed colonial and postcolonial countries "in the past" rather than as part of the present economic system. Based on influential works like W. W. Rostow's *The Stages of Economic Growth*, Africa is defined as being in the traditional space "set in opposition to Western modernity" and influenced the Kennedy administration's "liberal Messianism" (Dunn 2003: 97 see also Easterly 2006: 24–25). Of course, this linear view of economic development completely ignores the fact that neither colonial economies nor postcolonial economies were national but rather extremely international, and linked up with the western economies that are defined as being in the "now."

In my frame analysis of Australia's *Foreign Correspondent* program I found that three of the 11 Africa stories used the big-man frame. The strongest use of big-man framing came in a report tellingly titled "The General's Dilemma." The ABC's own description of this report represents the big-man frame perfectly:

Nigerian General Martin Luther Agwai has taken on a United Nations-led Mission Impossible—making peace in the vast Darfur region of western Sudan region where conflict rages between the Sudanese army and rebel groups. ("The General's Dilemma" 2008)

In the report the notion of the strong leader trying to salvage a chaotic situation is used as the main narrative frame. The other two *Foreign Correspondent* stories about Africa that utilize the "strong leader" frame are stories about saving animals. The "strong leader" frame appears in subplots in the documentaries. In both cases we see strong local men helping international activists to stop the poaching of, in one case, bonobos ("The Swingers" 2010) and, in the other case, elephants ("Where Have All the Elephants Gone" 2013). In both cases the animals are portrayed as verging on the sacred and incredibly important to the future of humanity.[5] Thus, the poaching by the village-based locals is seen as both immoral (evil) and irrational in the long run, despite being a rational decision in the short term. The poaching is portrayed as being rational in the short term because it is driven by chaotic, poverty-stricken lives—and this is where the big-man frame comes in. The animal activists are local saviors, bringing the villagers resources in the form of jobs as rangers and also clinics and microcredit schemes. Thus, these two cases employ various elements of the tropes and frames we have been discussing: a situation is portrayed as chaotic, destructive, immoral, and outdated, but a strong local leader is found who may eventually drag people, despite themselves, and with western support, into modernity.

Victim Frame

If Africa, in journalism, is a "hopeless continent" run over by "the four Ds of the African Apocalypse—death, disaster, disease, and despair" (Hunter-Gault 2006: 72), it is also a continent peopled by victims of these apocalyptic horsemen. Journalist Fergal Keane writes in a critical essay: "In this continent the locals exist in a state of perpetual famine, corruption, disease and warfare" (2004: 8–9). The key word in Keane's quote is "perpetual," and there are two problems with this consistent portrayal of victims. One is overuse and the danger of this being the *single story* of Africa (Adiche 2009). The second problem is in the mode of representation, a mode I will discuss as "poverty porn."

There is evidence that the victim frame is overused to the point of skewing perceptions of reality, and of becoming a stereotype by presenting a type of generic victim image. In terms of misrepresenting reality, in 2011 the *Columbia Journalism Review* published research that found that "between

May and September 2010 the ten most-read US newspapers and magazines carried 245 articles mentioning poverty in Africa, but only five mentioning gross domestic product growth" (Rothmyer 2011: 18). This is despite the fact that, at the time, poverty rates were falling steadily and more quickly than expected, and African economies were in general out-performing western economies in terms of GDP growth (2011: 18).

This overpresentation of Africa and Africans as victims has been criticized in the world of INGOs for some time. In 1988, at least two separate reports and a conference dealt with this issue in the UK (Coulter 1989: 12), and in 1989 the General Assembly of European NGOs adopted a Code of Conduct on Images and Messages Relating to the Third World, which among other things aimed to move away from fatalistic images of victims (Benthall 2010: 182–84). Images of starvation and disaster were found to have created a skewed view among teenage survey respondents, who saw African countries as hopeless and incapable of helping themselves. These attitudes crossed over to impact minorities living in the west, with teenagers from these groups recounting instances of racist and patronizing attitudes toward themselves or, in other cases, describing how they had been influenced by the media into believing themselves inferior (Coulter 1989: 12).

These teenagers were impacted by the images that had been overused to the point of becoming stereo*types*. Two such types are the "Biafra child" and the "refugee." In both cases the use of these stereotypes removes specificity and flattens context.

Until the 1980s the common aid-appeal images of helpless passive victims were known as "Biafra children" (Paech 2004: 23). Paech argues that "the perception of Africans as helpless victims is notoriously emphasized in images of anonymous abandoned children. . . . Nameless hungry children become generic icons that can be shot to order" (2004: 20). In evidence of this she presents the case of photojournalist Kevin Carter's famous, harrowing image of a vulture waiting for the death of a Sudanese child. This image was purchased by Save the Children and used in advertising despite the organization not actually operating in Sudan at that time (2004: 24). The specificity of this child's suffering becomes irrelevant; only the stereotype of sub-Saharan victimhood remains.

Competing with the Biafra child for the stereotypical victim image of Africa is that of the refugee. In the highly regarded satirical essay "How to Write about Africa," Wainaina makes specific mention of the refugee image:

Among your characters you must always include The Starving African, who wanders the refugee camp nearly naked, and waits for the benevolence of the West. Her children have flies on their eyelids and pot bellies, and her breasts are flat and empty. . . . She must look utterly helpless. She can have no past, no history; such diversions ruin the dramatic moment. Moans are good. She must never say anything about herself in the dialogue except to speak of her (unspeakable) suffering. (2005: 93)

Mbembe, in his 2001 book, similarly highlights the stereotype of the refugee, but he does not need satire to draw it out—he simply quotes from a *Le Figaro* account of Rwandan refugees returning from the Congo. The report draws on images of Dante, humans "melting together in astonishing uniformity" and "nothingness"; as described by Mbembe, it is a "vast scaffolding of dead elements."

This is the kind of mirror held up before the continent at the end of a frenzied century. What do we see in it? A brief and dissipated life in every sense. Men and women who pass by and change, forms, languages, animal figures deprived of sound. The spectacle of the world marked by unbridled license. (Mbembe 2001: 238)

Notice that evil, apocalyptic visions are brought to mind by the reference to Dante; there are animal images, a lack of voice and knowability, and the fact that all the individual people have been turned into one, an almost natural, primal backdrop. This ability of the victim frame to incorporate so many of the tropes of the Africanist discourse needs to be highlighted, for it is a deeply problematic issue. Mamdani makes the point that the victim frame reinforces the African as primordial and apolitical: "Journalism gives us a simple moral world, where a group of perpetrators face a group of victims, but where neither history nor motivation is thinkable because both are outside history and context" (2007: online).

I considered the continued relevance of these critiques—and how they might link to the current development discourse of bare life, posited by Duffield, when analyzing a scene in an episode of *Foreign Correspondent* in which a journalist stops off in an IDP camp during her journey to meet a group of militia in the Darfur. The first image of people that we see in the camp is seemingly abandoned children in the trope of "Biafra": two preteens, in a wide shot that makes them seem even younger, with a stark, empty desert scrub background, are somehow managing to survive. One is carrying fire-wood and one a baby sibling. This is followed by an interview with a mother in the camp. It is conducted by the journalist, Nima Elbagir, who is Sudanese born, but western educated. In the interview Elbagir avoids many practices that Wainaina critiques. She gives us the mother's name and asks questions, trying to elicit Fatima's history before the camp. The translation that results is this: "She says their life used to be so good. They lived in brick houses. She said they had donkeys and livestock and she said *we had a means of living*" ("Meet the Janjaweed" 2008; my italics). It's obvious that this woman, before the war, was a pastoralist, potentially a moderately wealthy one, and I can't help asking myself why Elbagir didn't describe her as a pastoralist. Why emphasize the bare necessity of self-sufficient life? The answer could be because the media frame of victimhood has taught us to think bare life is enough for sub-Saharan Africans.

Poverty Porn Frame

The term "poverty porn" is widely referenced in the aid world and will come as no shock to people familiar with the cynicism that is "rife in development circles" (Wallace, Bornstein, and Chapman 2007: 175). I have adopted it because it is a useful metaphor for focusing on *how* victimhood in Africa is presented. Porn emphasizes brutal realism, shock, and the activation of visceral senses. There is a lack of respect in the way the actors/subjects are presented for consumption. Furthermore, despite the emphasis on the "realism" of the production, porn does not provide any sense of narrative "truth." By this I mean that it lacks the sense of a deeper understanding about the world that we might expect from other forms of storytelling.

In critiquing the manipulation of the victim frame by western journalists, a common example is "a photograph (often reproduced since) by Wendy Wallace of news photographers lining up to photograph a single starving child" (Benthall 2010: 180). Even more shocking than the poor child's exposure to a media pack is that the emaciated child had been brought out to sit in the dirt specifically so the TV crews and photojournalists could capture that image (Paech 2004: 19). Thus, we see a lack of respect both for the truth of the subject matter and for the subject.

The porn descriptor also applies because of what is considered acceptable for presentation. The conventions of presenting African poverty and distress allow images that are far more graphic than those applied to western poverty and distress. Rothmyer describes as "poverty porn" a June 2010 *Time* article that showed "graphic pictures of a naked woman dying in childbirth" (2011: 18). Eriksson Baaz and Stern similarly describe "the ways in which outsiders have rendered [rape] survivors' testimonies" as having "frequently been characterised by a pornography of violence. . . . The often intimate representations of injured bodies and suffering are composed in a way that would be quite unthinkable if those depicted were survivors of sexual violence in most countries in Europe and the USA" (Eriksson Baaz and Stern 2013: 92).

The authors go on to point out the different standards of practice for journalists and INGO workers when at home compared to overseas:

> Who would even ponder the idea of letting journalists and other visitors into a hospital ward in New York or Stockholm with women waiting for, or just recovering from, surgery for rape-induced genital injuries, and urge them to speak and retell their stories to complete strangers? . . . [In contrast,] Congolese women appear as different; as women who are there "to be seen" . . .: who do not have to be protected from reliving the traumas of rape by retelling their stories over and over again. They emerge as the visitors' "private zoo[s]" . . . ; as objects whose sufferings are there to be consumed by a Western audience. (92)

Even *on the day of writing this*, on October 22, 2015, another example of the cavalier treatment of an African female rape victim is playing out in the media. The case concerns a Somali refugee caught in Australia's disgraceful asylum seeker system and held on the small island nation of Nauru, where she was raped. She was flown from there to Australia to receive treatment for the resulting pregnancy but was returned to Nauru without treatment. The associate editor for *The Australian* newspaper, Chris Kenny, said to be the first foreign journalist granted a visa for Nauru in over a year, traveled there to interview the 23-year-old woman, known as "Abyan." He is accused of bringing Abyan to tears and questioning her testimony. The refugee advocate Pamela Curr notes: "This would never happen to an Australian woman. There'd be an outcry. You can just imagine a journalist and police arriving on the doorstep of an Australian woman who'd been raped. It's so indecent" (Robin 2015: online).

In this context of poverty porn it is no surprise that the cofounder of humanitarian organization Médecines Sans Frontières (MSF), Xavier

Emmanuelli, recognizes "that MSF owes its success to cooperation with the media, yet he [Emmanuelli] *despises* television and in particular the reporters who zoom in on the sufferings of victims to give viewers the modern equivalent of the Roman circus" (Benthall 2010: 135).

The modern equivalent of the Roman circus. It's a description that bears repeating.

The White Hero / White Man's Burden Trope

> The great explorers' writings are nonfictional quest romances in which the hero-authors struggle through enchanted, bedevilled lands toward an ostensible goal: the discovery of the Nile's sources, the conversion of the cannibals. But that goal is also sheer survival and return home to the regions of light. The humble but heroic authors move from adventure to adventure against a dark, infernal backdrop where there are no other characters of equal stature, only bewitched or demonic savages.
>
> —Brantlinger 1988: 180–81

In the—what shall we call them, semifictional?—works of the "great explorers," the white hero trope rests upon many other tropes of Africa, including the lacuna and savagery tropes. And yet this trope continues to find its way into adventure stories for the teenage market set in Africa (Logan 2001), and those original colonialists are still celebrated as explorers and heroes, a celebration that is insulting when one considers what their "heroism" is defined in opposition to. For this reason, at the beginning of 2015, students fought for the removal of a statue of Cecil Rhodes from their South African university campus with the slogan "All Rhodes lead to the colonisation of the mind" (Davis 2015: online).

The white man's burden refers to the title of Rudyard Kipling's poem welcoming the United States into the "civilising mission" of the west through that country's colonization of the Philippines. The poem plays on the white hero trope already created through colonial adventure stories, and it elevates colonialists to an elite stratum, for in Kipling's poem mothers were to send forth only "the best ye breed." However, with the word "burden" it also introduced a sense of noble self-sacrifice that was not so prominent in the original, adventurous white hero trope.

The fact that this poem was written in 1899 after some of the worst travesties of empire had been exposed, including Stanley's outright lies,[6] seems incredible. However, the Africa of colonial heroes was too strongly ingrained to lose ground, so those who went forth to undertake the white man's burden as missionaries or civil servants were also elevated as heroes, and this has spawned as great many variations in modern media frames.

Angel of Mercy Frame

> No piece from an Africa disaster zone is complete without the sound bite from a white angel of mercy from one aid agency or another.
> —Keane 2004: 9

The continued life of the "white hero with a burden" trope in the "white angel of mercy" frame is obvious. The general public, consuming foreign correspondence, is given to understand that Africans are sitting there waiting to be helped, that "the only healthy, happy people are 'aided' people" (Coulter 1989: 12). Journalists feed this belief partly as a result of the reporting norm of seeking "relate-ability." Keane describes how journalists use the "white angel of mercy" in stories because they think familiar faces will help viewers relate to the story:

> Rarely do TV journalists pause to contemplate the consequences of this color-coded compassion. Viewers at home are watching (usually) a white reporter and white aid worker, and beyond them almost as backdrops are the wretched African masses. Just as it's always been and always will be, they think. Thank goodness for our brave reporters and aid workers. (Keane 2004: 9)

Keane's point is that the motivations for including the "white angel of mercy" may be "good" but the outcomes are problematic. Unfortunately, journalists are in the complex position of balancing pros and cons. Sometimes the inclusion of the white angel of mercy frame may be the difference between getting the story published or not. This is especially the case for long-running stories that have been forgotten or deemed un-important/unnewsworthy. In such cases the white angel of mercy frame becomes a news hook.

Peter Gill of Thames Television wanted to film famine in Karamoja, Uganda, in 1983. He was only able to do so by showing the arrival of five pretty, white nurses. Much the same "Angel of Mercy" approach was adopted by the British tabloids when the 1984 Ethiopian famine finally became a major story. (Harrison and Palmer 1986: 248)

In some cases the media can hardly be blamed for reinforcing the white angel of mercy frame. For example, in 1994 when Goma was inundated with both genuine refugees fleeing uncertainty, and *génocidaires* fleeing retribution, the town was deliberately plastered with INGO logos by the INGOs so that TV cameramen *couldn't* avoid getting them in their pictures when crafting their stories (Dowden 2009: 248).

White Hero Frame

What about the adventure side of the white man's burden trope? The Rhodes element? Has this found its way into the modernized white hero frame? Indeed it has. Percy Hintzen has identified the fusion of white compassion and adventure in his analysis of the 2007 special edition of *Vanity Fair* dedicated to Africa and guest-edited by Irish superstar frontman of band U2 and poverty campaigner Bono. Bono takes the opportunity to describe the creation of his social enterprise/charity, Product Red:

We needed help in describing the continent of Africa as an opportunity, *as an adventure*, not a burden. Our habit—and we have to kick it—is to reduce this mesmerizing, entrepreneurial, dynamic continent of fifty-three diverse countries to a hopeless deathbed of war, disease, and corruption. (Hintzen 2008: 78; my italics)

His guest editorial is paired in the magazine by a facing full-page ad for Dolce and Gabbana. It is a sexually provocative ad with the suggestion of a nude white woman being pleasured by two ambiguous brown bodies. Hintzen, who is particularly focused on the sexuality trope, describes the situation beautifully when he writes:

I was struck, but certainly not surprised, by the seamlessness with which all the tropes and figurative representations of white supremacy, (cross)racial desire, sexism, developmental historicism, and materialism combined in the juxtaposition of the two-sentence Bono quotation and the advertisement. (Hintzen 2008: 78)

Foreign Correspondents' "Dangerous Journey" Frame

The white hero trope also lives on when foreign correspondents emphasize their dangerous journeys. In speaking of the devastating, highly complex conflict in northern Uganda, Allen and Vlassenroot note that international news reporters have repeatedly focused on "telling adventure stories about their attempts to interview [Joseph] Kony" (2010: 3). The replacement of a story of complex conflict with a white (wo)man's adventure is another example of erasure journalism and the lacuna trope's foundational force.

Liebes and Kampf (2009) argue "performance journalism" is common to TV war coverage. In her analysis of Anglophone war coverage after 9/11, Palmer (2018) found that war correspondents tend to "become the story," and this had impacts on understanding the situation at hand:

> The news narratives of this era often focused on individual war correspondents' safety catastrophes. These catastrophes were transformed into melodramas that presented the war correspondents as the neoliberal subjects of late capitalism, while simplifying the deep structural problems that engendered correspondents' mishaps—and the geopolitical conflicts they were covering—in the first place. (Palmer 2018: 3)

Harris and Williams (2018: 100) similarly argue that the embedding of journalists with military units during the Iraq war created dramatic footage that "tended to exclude more sober analysis."

This discussion around journalism practice and "star reporters" will be further explored in Chapter 4. For now, the point to make is that these wider sociocultural and industrial forces interact with the Africanist discourse reinforcing its continuance. It is particularly important to acknowledge the discursive effects of this "dangerous journey" frame because the tendency to place the foreign correspondent at the center of the story could become a type of brand positioning in the increasingly competitive world of media content. This is a world where, especially online, emotionalized and personalized stories suit algorithmic preferences (Wilding et al. 2018: 37–39).

In 2011 *Foreign Correspondent* changed the opening scenes for all stories to focus on the foreign correspondent being on location. In 2012 these opening scenes were scrapped but a new overall introduction to the program was created. No longer did the program start with an image of the globe; this was replaced with a mini storyline of the foreign correspondent getting ready for the journey and boarding a plane. The globe image emphasized the place being traveled to; the new intros emphasize the traveler—the modern-day, Rhodes-inspired adventurer.

The Transactional Frame and the Closely Related Frame of Uber-westerner and Uber-victim

As mentioned earlier, the INGO industry has long been concerned with the effect of its victim imagery on public perceptions of developing countries. More recently, that concern has focused on the problem of what has been called the "transactional frame," a frame summarized as "People are in need, give money." This shift in focus was brought about in 2011 thanks to influential research by Oxfam and the UK Department for International Development.[7] That research finds "even engaged people can't sustain a conversation about debt, trade or aid for long" (Darnton and Kirk 2011: 6), which seems to clearly point to the influence of the lacuna trope. Reading the results of this research, one can see the potential impact of the tribal/irrational/savage trope in that people engaged with development are actually *more* likely than the general population to view poverty as a result of having corrupt leaders (Darnton and Kirk 2011: 22). The potential impact of the white man's burden trope is seen in the finding that the dominant paradigm of engagement for the public is "characterised by the relationship of 'Powerful Giver' and 'Grateful Receiver'" (Darnton and Kirk 2011: 6). This last element is crucial for Darnton and Kirk, who argue that the "transactional frame" actually fostered a *shallow engagement* between the givers and the people in crisis.

This shallow engagement effect is in line with the hydraulic functioning of frames, where attention given to one aspect drives out other possibilities. As Schön and Rein point out: "We cannot see the familiar Gestalt figure at one and the same time as both the old woman and a young one" (1994: 29). In the transactional frame the "give money" motif drives out other messages. In the "Make Poverty History" campaign:

The transformative potential offered by the rallying cry of "justice not charity" went unheard, in part because it was unfamiliar and hard to comprehend, and also because it was drowned out by the noise of celebrities, white wristbands and pop concerts. (Darnton and Kirk 2011: 6)

In a symbiotic relationship, as the Make Poverty History campaign became more closely associated with charity, it also became "Africanized." For a variety of reasons the worldwide campaign became particularly associated with Africa, with the result that "the strong justice agenda that underpinned the campaign at the start became incrementally replaced by an 'empathy' agenda" (Harrison 2010: 392). Harrison posits that this replacement was due to the British cultural history that sees Africa as "defined by poverty and subsequent empathy" (2010: 398). Darnton and Kirk agree that the transactional frame is particularly associated with Africa because "Africa is so bound up with charity in the public's mind" (2011: 33). They are less clear on the role of *empathy* in the association with Africa and in the transactional frame.

It seems contradictory to combine empathy with the transactional frame when we have already stated that the transactional frame leads to shallow engagement with issues of poverty. However, deep *engagement* with root issues is not necessary to feel *empathy*; on the contrary the transactional frame emphasizes the alleviation of suffering simply by making a small donation. Other recent framing research, which similarly focuses on the role of empathy in media images of "poverty and helping," has likewise found that "the 'spectacle' [of the poor] does not ask viewers to *engage* [with the poor person or the context for poverty] but only to accept the given domain [of caring for the poor] on which it rests" (Lancione 2014: 707; my italics). In the United States it is found that humanitarian and human rights-focused reporting is the focus of newspapers and TV, and this sidelines the potential to talk about economic justice (Powers 2018: 74–81).

This issue of engagement versus empathy is particularly important for journalists. There is no doubt that empathy is a fantastic narrative tool that can be used in a transactional frame to foster donations, but it is also true that the transactional frame removes context, and donations are not a long-term solution. As has been eloquently put by others: "By suggesting that their rural women beneficiaries are just one goat, or one $10 loan away from escaping poverty, the marketing messages of NGOs constantly do violence to the true sources of world poverty" (Barry-Shaw, Engler, and Oja Jay 2012: 53). Even outside of poverty, in situations of crisis, for example after environmental megaevents, the transactional frame does nothing to acknowledge the underlying social inequality attached to the likelihood of experiencing, and experiencing more severe, effects from these globally generated risks.

By showing "white angels of mercy" and giving donation details in their stories, journalists support this misleading transactional frame. And yet the temptation to use the transactional frame is strong because it supplies a clear role for both journalists and audiences in what are often overwhelming situations. As described by foreign correspondent Linda Polman: "When a message is painful, we always want a quick, ready-made solution to numb the pain" (2010: 172). African foreign correspondents speak with pride of how their stories have directly led to huge donations, for example $10,000 worth of drugs being donated in Sudan after a "Mother Teresa of Africa" story (Brill in Leith 2004: 56) and £70,000 being donated to Hospice Uganda to help AIDS sufferers after only one story in *The Telegraph* (Tweedie in Leith 2004: 369). Yet they also acknowledge this makes them "a cog in the world's humanitarian machine" (Balzar in Hannerz 2004: 46).

The role of foreign correspondents in the economics of humanitarianism is well understood by INGOs, which is why they increasingly fund overseas reporting trips for cash-strapped media companies (Polman 2010: 43), and this has framing effects. Powers (2018) argues that INGOs have been forced to conform to news values and styles so as to get their message into the news agenda and, at the same time, catch the attention of their existing and potential donors. However, the framing effects go both ways, especially on the ground in borderlands. According to Polman (2010: 44), "Having an aid organization guide you through a crisis zone is like looking at Europe through the eyes of the Salvation Army." Similarly Wright discusses how INGOs are able to "exercise significant amounts of interpretative influence through seemingly casual conversations about practical arrangements, some of which were held on planes or in jeeps, whilst accompanying freelancers to field sites" (2018: 259). Furthermore, traveling through scenes of chaos and devastation provides a "collective experience" (Tumber 2013: 54) somewhat analogous to that of war correspondents who go to the front line embedded in particular military outfits. It is no surprise that a similar result of identification with their guides (and protectors), and similar self-censorship, takes place.

In war reporting, the self-censorship revolves around not wanting to "embarrass" the soldiers by describing mistakes or confusion about their role (Ward in Tumber 2013: 57). Similarly, Hannerz found journalists chose not to publish stories about aid organizations misusing their funding for fear it will stop people donating (2004: 46). More significantly for audiences who rely on foreign correspondents to understand world events, self-censorship also leads to misrepresenting conflicts, as the correspondents edit out complexity so as not to interfere with their transactional framing. It has been widely noted that journalists misrepresented the post-Rwanda-genocide refugee crisis in Goma, failing to mention that the refugee camps also housed *génocidaires* (those involved in perpetrating the genocide) (Carruthers 2004: 164). Richard Dowden, former Africa editor for *The Economist*, and current director of the Royal African Society, when interviewed by the author in 2014, admitted to self-censorship in Goma, at least to the extent of downplaying complexity, because he did not want to stop people from donating money to help the refugees. He also said that, in retrospect, it was a mistake on his part.

Self-censorship has also been admitted to by those covering the 1980s Ethiopian famine (Polman 2010: 125), and again the shallow transactional frame misrepresented the truth. The famine was presented as an "act of God" caused by the weather. However, the truth of the matter is that weather causes droughts, not *famine*. The latter is a complex issue where politics plays a major role. In Ethiopia, these political causes were drowned out by the transactional frame (Franks 2013: 110; de Waal 1990: 79–98; Rothmyer 2011: 18).

Whether this failure to report the politics of the situation was justified is difficult to argue with certainty: "Even today, most journalists and aid managers take the view that it was in the greater interest of humanity to keep the issues simple and elicit the largest possible response" (Vaux in Franks 2013: 101). However, others argue that "over-hyped naively 'humanitarian' reporting can be as bad as no reporting at all" (de Waal 1990: 98). This is because depoliticized responses helped prolong the wars in Ethiopia and Sudan, with food supplied by the west for the famine victims going to soldiers instead. In southern Sudan, it was not until "Operation Lifeline" had run for 10 years that journalists started to question whether aid was prolonging the war, and received the answer that yes it was (Polman 2010: 129). Polman argues that foreign correspondents need to do a much better job of asking these questions sooner rather than simply reverting to the transactional frame.

Finally, the transactional frame is appealing to journalists because it has become deeply entwined with celebrity. To return to the Make Poverty History campaign, one of the ways it was "Africanized" was through the involvement of celebrities. The Band Aid concerts in 1985 developed a strong link between celebrities and Africa, and this has created its own easy-write frame (Rothmyer 2011). I have chosen to call this the Uber-westerner, rather than celebrity, frame to underline the echoes of the Africanist discourse. Nevertheless, the celebrities do create their own effects, and so they are not *just* a different version of white angels. In particular, celebrities reinforce the transactional frame through their links to consumer culture (Darnton and Kirk 2011: 107)

The Development Promise

It's time to bring the development promise to the foreground in our analysis and consider some difficult questions. First and foremost, why shouldn't journalists include the idea of development in their stories of sub-Saharan Africa? Surely the answer is yes when we consider the need for education, health, food, those common aspirations of humanity. Surely these wishes should *not* be absent from our stories?

Unfortunately, as Hall (2013b: 17–18) noted, to say something you must enter into language, and evoking the idea of development evokes its problematic baggage, such as Rostow's stages of economic growth that place colonial and postcolonial countries "in the past" rather than in the present economic system. Wolf makes a similar argument in his groundbreaking *Europe and the People without History* when he criticizes modernization theory for "casting such different entities as China, Albania, Paraguay, Cuba, and Tanzania into the hopper of traditional society, [which] precluded any study of their significant differences . . . [and] of issues demonstrably agitating the real world" (1982: 12–13).

The economic realities of the DRC are absolutely of our time, created by the economic system of today. If anything, the DRC is not a vision of our past but instead is the canary in the mine showing us our future. In his much-acclaimed history book David Van Reybrouck points out that the DRC has been at the forefront of economic and political development. The DRC has been at the forefront of international trade for

> ivory in the Victorian era; rubber after the invention of the inflatable tire; copper during the full-out industrial and military expansion; uranium during the Cold War; alternative electrical energy during the oil crisis of the 1970s; coltan in the age of portable telephonics. (2014: 119)

This has then connected to, or created, the politics of significant world developments. The first American atomic bomb was made with uranium from the DRC (Van Reybrouck 2014: 190); the speech of the first president of the DRC, Patrice Lumumba, at the proclamation of Congolese independence, is considered one of the great speeches of the 20th century and a key text of African decolonization (2014: 272); Lumumba was responsible for opening up the Cold War front in Africa (2014: 298) and was an early victim among CIA-sponsored assassinations (2014: 310). The second United Nations secretary general, Dag Hammarskjöld also lost his life in the DRC; the exact why and how remain unclear (2014: 281) but perhaps because he got in the way of multinational mining companies (Muehlenbeck 2012: online para. 13). Van Reybrouck goes on to describe the current violence in the DRC as

> no atavism, no primitive reflex, but the logical result of the scarcity of land in a wartime economy in the service of globalization—and in that sense, a foreshadowing of what is in store for an overpopulated planet. Congo does not lag behind the course of history, but runs out in front. (2014: 471)

The word "development" erases all this and instead gives the false impression that what has happened so far *and the politics of today* are unimportant because people are still "developing" and not fully fledged world players. I would even argue that it erases the need to really care about what is going on; for all we need to do is wait and they will catch up. The word "development" also erases specific political context. Duffield writes: "In separating life from politics—by holding it above the fray of battle in the name of neutrality—humanitarian emergency strips away the history, culture and identity of the peoples concerned" (2007: 33–34).

Again, the DRC is a perfect exemplar of this removal of politics. When the Rwanda genocide led to a mass exodus into Zaire, "Given western conceptions of refugees, the international humanitarian community treated these individuals as victims, regardless of their role in the 1994 genocide" (Dunn 2003: 151). This conceptual move by the international humanitarian community (which, as mentioned already, was followed by the foreign correspondents) had serious repercussions that are still being felt today as the Rwandan invasion of Zaire was justified on the grounds of the need to destroy the *génocidaires'* stronghold, built up in the refugee camps, under the eyes of the international community. Rwandan president Paul Kagame argued that the Interahamwe (a Hutu paramilitary organization) had control of the camps and were regrouping. The battle to oust them helped establish the east of the DRC as a permanent war zone run by militias from many different countries who took over different mines and other sources of income.

The second part of Duffield's argument is that a certain type of development follows humanitarian emergencies, and this argument is also relevant to the question of why we should be wary of uncritically including the "development promise" in stories. Duffield argues that the depoliticizing effect of humanitarian emergency prepares the ground and "the foot soldiers of development follow behind, rebuilding communities and promoting small-scale ownership of property in the interests of improved self-reliance" (2007: 33–34). At this particular historical juncture the "development promise" implies specific actions delivered through the current global aid structure, actions that are, at the very least, not as unproblematic and benevolent as is generally assumed in the media. If the delivery of humanitarian aid through these global structures was unproblematically beneficial, why did the Australian prime minister Kevin Rudd refuse to accept foreign aid after the devastating 2011 floods? Rudd argued that allowing aid agencies in to help was one of worst things Australia could do. Describing this incident, Ramalingam notes that "Rudd, incidentally, had previously overseen a promise to double the Australian foreign assistance budget, aid presumably being among those things which one is more blessed to give than to receive" (2013: 4). Duffield problematizes the issue further by arguing:

> Development shares with liberalism an experience of life that is culturally different as always being somehow incomplete or lacking . . . this impoverished experience of life, and its accompanying will to exercise moral tutelage, is an enduring feature of liberal imperialism. (2007: 216)

Duffield also argues that aid and development structures, since decolonization, have been creating a deepening "biopolitical division of the world of peoples into developed and underdeveloped species-life" (2007: ix), shaping a terrain of unending war.

Through the call to action, the transactional frame satisfies the journalism prerequisite of immediacy, and this fact tracks well in the humanitarian field. Policy, communications, and marketing personnel know that the way to attract journalists to a long-running, or underreported, emergency is by "ramping up" the need for action (as the director of a major INGO in

Australia told me in 2013). INGOs, through their communication tools, create a sense of movement, drama, and urgency to help journalists "see" the story. However, the complexity of conflicts means aid money may in fact end up being used as a weapon that prolongs war (Polman 2010).

> Critical observers would even argue that in instances of civil conflict, an intricate, partly symbiotic structure of relationships could be set up between strong international NGOs, weak government, rapacious warlords, news media and benevolent metropolitan audiences. And it might end up mostly serving the material interests of the warlords. (Hannerz 2004: 142)

This technique of "ramping up" to attract attention is probably also understood by warlords, who, instead of using communication tools to create a greater *representation* of emergency, use their weapons to *actually* create a greater emergency. Seaton (1999: 59) contends that "warring groups all over the world now seem to understand how cynically to exploit their own civilian suffering and the frailty of the position of peace-keeping forces in order to manipulate the media." Polman recounts a shocking interview with Sierra Leone Revolutionary United Front (RUF) rebel leader Mike Lamin in which he tells her "you people" weren't interested in the civil war in Sierra Leone, only "the white man's war in Yugoslavia and the camps in Goma," and so the RUF consciously ramped up the violence. Lamin goes on to say, "It was only when you saw ever more amputees that you started paying attention to our fate" (Polman 2010: 167).

This, of course, does not mean that all aid relief and development projects are unhelpful, or that people working in development are ignorant or evil. It is important to honor the needs and wishes of people to achieve such basic rights as education. However, it does mean that, as journalists, we should move away from uncritically incorporating the *traditional* idea of the "development promise" into our stories. In a world where "every country on earth is part of the aid system, as donor or recipient, or increasingly both" (Ramalingam 2013: 4) I am arguing that development needs to be taken out of its colonial white man's burden frame—our reporting on development needs to recognize the complexity of the system and our frames need to be updated.

This is absolutely possible. Foreign correspondents have written books based on extended research for the purpose of usurping, or going beyond, the media stereotypes of Africa that the Africanist discourse has helped create.

These books include *Hotel Africa* (Zachary 2012), *Africa, Altered States, Ordinary Miracles* (Dowden 2009), *Gogo Mama* (Sara 2007), *New News out of Africa* (Hunter-Gault 2006), *A Continent for the Taking* (French 2005), and *Africa, Dispatches from a Fragile Continent* (Harden 1991).

Apart from these works, consciously sought out by me, on any given day, as part of my normal media consumption, I may come upon an article from sub-Saharan Africa that does a brilliant job of presenting complexity and avoiding the frames associated with the Africanist discourse. Like the day when the *Global Post* ran a story titled "It's April, Which Means Eritrea's Refugees Are Headed North" (Belloni 2014). It was a superb piece of journalism that presented refugees making the dangerous journey to Europe, over desert and sea, not as irrational criminals, nor helpless victims manipulated by people smugglers, but as rational actors in a complex, uncertain and unfair world.

My literature review has also thrown up surprising knowledge. For example, the biennial charity appeal in the UK known as Comic Relief would, one would naturally assume, fall into the transactional frame of shallow engagement. However, Benthall argues that these comedians have been able to create a more meaningful media frame by making local people laugh, making them seem real and close to the western viewer (2010: 87–88).

That there are alternatives to the Africanist discourse is important. We should not lose sight of them, but the fact of their existence does not nullify the overall argument that Africanist discourse continues to live in modern media frames. Nor does it lessen our need to understand how the power of Africanist discourse affects our practice, in terms of writing styles and beyond that in terms of our actions. To begin considering the latter element, I wish to posit the existence of something I have termed the "lacuna effect."

The Lacuna Effect

The tropes of colonialism have discursive effects that live on. The white man's burden effect has been documented in the aid industry, most prominently by William Easterly in his book *The White Man's Burden: Why the West's Efforts to Aid the Rest Have Done So Much Ill and So Little Good*. In it Easterly writes:

> From the beginning, the interests of the poor got little weight compared with the vanity of the rich. The White Man's Burden emerged from the

West's self-pleasing fantasy that "we" were the chosen ones to save the Rest. The White Man offered himself as a starring role in an *ancien régime* version of Harry Potter. (2006: 23)

Dambisa Moyo in *Dead Aid* similarly argues that the problems with aid in Africa stem from a white man's burden mindset. She writes: "There is, of course, the largely unspoken and insidious view that the problem with Africa is Africans. . . . That, somehow, deeply embedded in their psyche is an inability to embrace development and improve their own lot in life without foreign guidance and help" (2009: 31). In international relations Koddenbrock (2012) has outlined how interventions in the DRC are based on a logic of "functional pathologization." This involves an assumption of Congolese deficiencies and the benefits and legitimacy of western intervention. Eriksonn Baaz and Stern point to the Congo as "once again . . . a site of European (and American) adventurism and benevolence." They then draw on Spivak to argue that "the massive engagement in the plights of Congolese rape survivors serves as an illuminating example of the re-enacting of the white wo/man's burden to 'sav[e] brown women from brown men'" (ErikssonBaaz and Stern 2013: 92). The point Easterly, Moyo, Koddenbrock, and Eriksson Baaz and Stern make is that they believe the preoccupation with the white wo/man's burden has a negative effect on the quality of aid in Africa.

In a vein parallel to the preceding analyses, my work has illuminated a negative lacuna effect on the quality of journalism in Africa. The lacuna effect on journalism is both a lack of knowledge of what exists and, more importantly, *an inability to understand that this knowledge is lacking.* After all, Africa is a place of nothingness, without history or political agency. A blankness of mind is therefore the norm. In the words of foreign correspondent Blaine Harden, Africa is a "gloomy question mark" (1991: 14). Howard W. French, who was West Africa correspondent for the *New York Times* from 1994 to 1998, in critiquing his own, and his colleagues', performances, seems to point directly at the influence of the lacuna. Speaking of raids on refugee camps in Zaire following the Rwanda genocide, French writes:

The scramble to do some rudimentary ethnic detective work brought to mind just how normal it was for reporters to operate in nearly perfect ignorance of their surroundings on this continent. Africa remained terra incognita for most within my profession, whose job it was to inform the world, and for many of us an assignment here involved little more preparation than thumbing through a *Lonely Planet* guide. Anywhere else in the world we would have been judged incompetent, but in Africa being able to get somewhere quickly and write colorful stories was qualification enough. It was a repeat performance of the same contemptuous glossing over that characterized so much of Europe's colonial involvement with the continent, and though I had more experience here than most of my peers, I was in no way exempt. Only midway through Kabila's campaign against Mobutu did I finally get around to reading *The Rise and Decline of the Zairian State*, Crawford Young and Thomas Turner's seminal 1985 study of Zairian politics and history, which should have been a prerequisite for any reporter. Scales fell from my eyes in the face of such detailed knowledge, and I felt a deep, physical sense of embarrassment at my own ignorance. (2005: 128–29)

Basil Davidson (a foreign correspondent and historian) also seems to point to the lacuna's blinding effect when he laments: "The political sociology of Africa, in brief, has been peculiar to itself but peculiar in no other sense. Its seeming eccentricity or inexplicability or unpredictability has existed only in the eyes of those who have not really looked" (1992: 63). He goes on to say that among the mistakes caused by this inability to overcome colonial topes was the mislabeling of the politics of clientelism as the politics of tribalism—with the real tragedy being that the politics of clientelism was a more divisive force than tribalism could ever be (1992: 206).

Davidson refers to the decolonization and Cold War years, but this mislabeling as a result of ignorance continues. Gruley and Duvall's analysis of the Sudan Darfur conflict uncovered the following example from the *Washington Post* (WP):

In one notable article, the WP's reporter Emily Wax sought to debunk five "misconceptions" about Darfur, including that "Africans" and "Arabs" can be distinguished by the color of their skin. She wrote, "Although the conflict has been framed as a battle between Arabs and black Africans, everyone in Darfur appears dark-skinned" (WP, 23 April 2006). As the WP's Nairobi bureau chief from 2002 to 2006, Wax had an important role in the initial reporting that framed the conflict in tribal terms, and several of her earlier articles represented categorical, physical differences between "Africans" and "Arabs" (e.g. WP, 27 June 2004; WP, 18 July 2004; WP, 29 September 2004). (Gruley and Duvall 2012: 40)

To what can we attribute this change of frame by Wax, other than the fact that her early ignorance was being corrected as the conflict dragged on? Unfortunately, she was in the final year of her four-year posting when she wrote the "debunking" article. One can only hope that the correspondent who replaced her would have read it and that this did some good—at least in the reporting of Darfur. But what about all the other regions that same correspondent would have to cover? Without an acknowledgment of the underlying problem—that is, the lacuna effect—this one corrective article is unlikely to have had a wider impact on this incoming journalist's reporting. This is the tragedy of the lacuna trope leading to a lacuna effect; it has the self-validating power to make Africanist discourse true because we don't realize that we're ignorant, and so we don't go looking for more information.

Easterly seems to point to the lacuna effect when he argues that

both the IMF and the World Bank produce reports on individual poor countries, which are available on their websites. Together these reports make up the world's best supply of information on the economic situation of individual countries—most of them are ignored by the American press. (2006: 189)

The evidence of a failure to look is damning enough, and yet there is another way to envisage the impact of the lacuna trope on the profession—that is, through lack, full stop. Besova and Cooley point out "coverage of the African continent produces only 5.6 per cent of international news produced by US media" (2009: 219). How can one account for journalism's willingness to accept this situation, other than with the fact that ignorance concerning Africa is the norm? Despite the way in which new technology has eased the transmission of information, "The lack of coverage of some of the world's most deadly and long-lasting conflicts, usually in Africa, remains a feature of digital reporting" (Williams 2019: 179).

In terms of a lack of reporting and a lack of understanding, the DRC is again an extreme example. Van Reybrouck points out that "since 1998 at least three million people and perhaps as many as five million people have been killed in hostilities in Congo alone, more than in the media-saturated conflicts in Bosnia, Iraq and Afghanistan put together" (2014: 400). He points out that the justification given for this lack of attention is that the war was incomprehensible and obscure. And yet, as he demonstrates, "A simple cartographic comic strip is all one needs to understand the course of events" (Van Reybrouck 2014: 441–42). McNulty (1999: 284) highlights the impact of tribalism framing in reinforcing the lacuna effect when he contends that by late 1994 the Rwandan genocide had been "filed away under 'inexplicable ethnic bloodbaths—Africa.'"

A lacuna effect is also obvious in the tendency described earlier of "erasure journalism," which removes Africans from their stories and replaces them with westerners. Take the case of the Ethiopian government's Relief and Rehabilitation Commission (RRC), set up in 1974 after a major famine, to try to prevent future famines. Harrison and Palmer write:

> Many argue that it [the RRC] is the best organisation of its kind on the African continent. It was certainly very largely responsible for the fact that severe drought in the south of the country in the late 1970s did not lead to greater casualties. The RRC is a prime example of an African organisation grappling with an African problem, year in, year out, with scant resources, in a quite heroic way. . . . In the countdown to disaster in 1984, the RRC's warnings, its forecasts and its estimates of need were all chillingly accurate. The UN, however, had its own, highly-paid "experts" on the spot. They reported differently, almost complacently. *The world listened to the foreign experts.* The world was wrong and Ethiopians died as a consequence. This shameful episode is fully documented in Peter Gill's excellent *A Year in the Death of Africa.* (1986: 98; my italics)

In the preceding excerpt we see a textbook example on the part of the development industry of a white man's burden effect in not trusting Africans to know their own best interests. However, I argue that journalists listening to foreign experts are influenced by both the white man's burden effect and the lacuna effect. I believe the lacuna effect is apparent in the failure to follow normal journalistic practice that would have dictated cultivating a strong relationship with the RRC as an important local source. A preexisting relationship would have meant the RRC's warnings would not have been ignored by journalists, even if these warnings were being contradicted by white others.

Finally, I wish to consider an observation that may help explain why we do not easily recognize the lacuna effect. David Newbury writes of sub-Saharan Africa:

This is a region not well known in the west, but one nonetheless enveloped in a century of powerful imagery-ranging from the "Heart of Darkness" to the "Noble Savage." In other words, it is an area that outsiders feel they "know" well. Consequently, these events have often been misunderstood— and the reporting on them has sometimes reinforced and extended the stereotypes which many outsiders carry about the people and cultures of this region (and of Africa as a whole). (1998: 76)

This statement is enlightening because Newbury draws out a distinction between familiarity and knowledge. As journalists, we may be *familiar* with the "gloomy question mark" (Harden 1991: 14), we may be *familiar* with *how to write about Africa*, but this doesn't mean that we *know* Africa. What a crazy statement anyway, to talk about "knowing" "Africa." As Dowden points out: "Africa has more than 2000 languages and cultures . . . in comparison, Europe is homogenous, America monotonous" (2009: 9).

Perhaps the simple and difficult truth is that to exorcise the lacuna effect, we have a hell of a lot of hard work to do. But this work must be done because right now the lacuna effect combines with simplified humanitarian perspectives to obscure the dynamics of borderlands. We must work to understand what we don't understand, gain knowledge, and make clever and complex representation decisions. It requires that we act as individuals within a system we are consciously trying to decolonize and that we resist our natural tendency to work as a cog in the machine. Only by calling out and removing the continued representational life of colonial narratives of Africa, and by acknowledging and fighting against the lacuna effect, can a new lexicon of understanding for borderlands be found.

This will not be easy. The words and concepts do not come easily to mind for the working journalist, nor will they be easily taken up by newsroom editors and western audiences, struggling to come to terms with a world risk society, who would prefer to believe in simplified, recognizable storylines of heroes and victims. Seaton, in critiquing the overuse of ethnicity as an explanation for contemporary war, argues that "journalism is the art of the cliché. Great journalism may inflect or bend the language for new purposes, but it is still dependent on the language and ideas that audiences recognize" (1999: 59).

So, to take up this antiracist challenge is to believe that "through the critical use of language we not only reclaim our shared histories, cultures and identities but develop a sense of belonging and connectedness to each other, a common language of understanding within and across difference" (Dei 2006: 29–30). *But how?* In Van Reybrouck's masterful history, he often confronts the tropes of the Africanist discourse head on. For example, when discussing the regional strongholds of different parties after the results of the Congo's first elections, he writes:

> The electoral map of Congo in 1960, therefore, was largely identical to the ethnographic maps drawn up by the scientists half a century before. This tribal reflex should not be seen as atavistic. Were pan-European elections to be held in Europe today, after all, there is a great chance that most of the French would vote for a Frenchman and most Bulgarians for a Bulgarian. In a vast country like the Congo, where the greatest part of the population had no more than a primary school education, it should come as no surprise that many voted for candidates from their own region. (2014: 264)

This strategy is interesting. It conflicts with the *Finding Frames* (Darnton and Kirk 2011) research. In *Finding Frames* Darnton and Kirk draw on the research of cognitive linguist George Lakoff to argue that certain words should be avoided because they conjure up powerful frames, and once that frame is conjured you cannot fight against it. For example, if you oppose reducing taxes but you describe that reduction as "tax relief," the strong frame around the word "relief" means you have lost the argument (Darnton and Kirk 2011: 73). Thus, Van Reybrouck may be doing a disservice to the point he is trying to make by using the words "tribal" and "atavistic." However, one could also argue that the tribal trope of Africa is so strong that once Van Reybrouck mentioned the regionally affiliated results, it would have been evoked anyway, so better to tackle it head on.

What Van Reybrouck also does is *reframe* through the process of metaphor "by which a familiar constellation of ideas is carried over (*meta-pherein*, in the Greek) to a new situation, with the result that both the familiar and the unfamiliar come to be seen in new ways" (Schön and Rein 1994: 26–27). By applying the results of the Congolese elections to regional European elections, the regional proclivity of both Congolese and Europeans come to be seen in new ways.

Schön and Rein (1994) argue for the power of these metaphors, and that finding the right metaphor, through a process of frame reflection and design rationality, can help resolve otherwise intractable policy controversies. In the next part of my research, the move from critique to the search for solutions, I was keen to see if Schön and Rein's theory, and methodology, could be transferred to journalism. I would test this out in Goma, DRC, reporting in, and around, refugee camps. It was time to start the experiment.

2

Writing with Interlocutors

Dateline: Goma, DRC, August 26, 2014

Five women: Rachel, Faulestine, Marie-Claire, Justine, Collette—these are my first interviewees in Goma and they are amazing. Each runs a grassroots civil society organization fighting the effects of the continuing conflict in Goma from the bottom up. Each has been personally impacted by the war and each now advocates for women's rights, both through their individual organizations and through collaborating to form a national body, Sauti Ya Akina Mama Mkongomani (hereafter Sauti), which they formed so as to have a greater say in regional, national, and international politics. I know all this, and yet I did not do the job I would have wished, both in terms of getting to know them and in terms of transmitting their story.

These women gave me far more than I gave them.

This chapter is about trying to apply theory to practice. The previous chapter pointed me toward the problems to be overcome, namely: the continued representational life of colonial narratives of Africa, and the lacuna effect, which combines with humanitarian perspectives to simplify and obscure borderlands. My first task was to tackle those "long dark shadows in my head" by introducing new frames of interpretation.

Ervin Goffman argues human beings are constantly asking: "What is it that is going on here?" (1975: 8) and we answer that question by accessing our primary frameworks. Those are the broad interpretive schema made up of relatively stable and socially shared category systems (Tewksbury and Scheufele 2009: 10). Journalists, as media producers who perpetuate existing media frames or introduce new ones, often fail to recognize that the schema (primary frameworks) we carry around in our heads are what we use to create those media frames. Moreover, particular frames speak to particular discourses to which we have granted "truth" status.[1] We fail to realize this because frames are "usually tacit, which means they are exempt from conscious attention and reasoning" (Schön and Rein 1994: 23). This meant that before I could attempt to find new frames for my reportage I needed to be exposed

Borderland. Chrisanthi Giotis, Oxford University Press. © Oxford University Press 2022.
DOI: 10.1093/oso/9780197565797.003.0003

to new, socially shared category systems and integrate these into my primary frameworks of interpretation; in short I needed to see the world from a different perspective. I needed to see through the eyes of those who have lived and understand borderlands.

But how does one develop new eyes?

Enter *Frame Reflection: Towards the Resolution of Intractable Policy Controversies*. Schön and Rein (1994) begin *Frame Reflection* with these words:

> This book is about the kinds of policy controversies that are all too familiar to readers of the daily newspapers and watchers of the evening television news—controversies about issues such as poverty, crime, environmental protection, the Third World, and abortion, which are highly resistant to resolution by appeal to evidence, research, or reasoned argument. (1994: xi)

The authors argue that some issues resist resolution in the public sphere because the antagonists are people using different interpretive frames. The authors believe it is only if those differing underlying frames are exposed, and discussed, that there will be a chance of moving the discussion. The process of exposing and co-creating new frames with interlocutors is called co-design.

I felt the appropriateness of using *Frame Reflection* as I knew I would be reporting from refugee camps and the issue of refugee movements is an intractable policy controversy. In Australia, in the midst of my preparation for the trip, Prime Minister Tony Abbott claimed election victory, propelled, in part, through his three-word slogan "Stop the boats." In the acrimonious debate around the safety of traveling to Australia by boat, a great many Australians were unable to comprehend people's need to take their fate into their own hands in terms of searching for a decent quality of life. This suggested a view of those people as belonging to a different species of humanity, a species able to live without dreams and hopes—a view that for some people bare life is enough.

An example of this "bare life is enough for refugees" view appears in the third season of *Go Back to Where You Came From*, a reality-TV series that sees Australians with differing views take the refugee journey to Australia in reverse (traveling backward from Australia through transit countries and then to war-torn points of origin). Melbourne teacher Andrew Jackson, initially anti-refugee, says of Iraqi asylum seekers stuck in stateless limbo in

Indonesia, that it had been "easy [for him] to say, oh you've come to Indonesia, it's an Islamic state and you've escaped being killed, so that's it, you've escaped. *But then you've got no life. And I've never worried about it from that point of view*" ("Go Back to Where You Came From S3 E2" 2015: @20m52s; my italics). It seems he had never thought about quality of life before, because he had never had the issue framed this way before. Perhaps this was because the issue was being framed for him, for the first time, by the asylum seekers themselves. It was this sort of expansion of my own primary frameworks of interpretation that I wanted to find, and so I decided that my co-designers had to be refugees. However, to be able to influence my primary frameworks, this work had to happen *before* the writing process and, as I would only have a 21-day visa for Goma, it made sense to attempt this process while still in Sydney with former refugees from the Democratic Republic of Congo now living in Australia.

To transfer the process of co-design described in *Frame Reflection* from policy formulation to the world of journalism, I needed to consider how I would mimic the process of situated frame reflection—that is the frame reflection that happens between interlocutors aided by discussion of materials. My solution was to give my interlocutors specific journalism examples to look at. I designed a semistructured interview process around three different media clips that were representative of commonly used frames for refugee stories in sub-Saharan Africa. Specifically, these clips made strong use of chaos, victim, and bare-life framing. I chose one-on-one interviews rather than any sort of group setting, as the one-on-one interview is the key research method of journalism and thus most easily integrated into existing journalism practice. Moreover, it has long been argued that the value of qualitative interviews is their ability "to capture how those being interviewed view their world, to learn *their terminology and judgments*, and to capture the complexities of their individual perceptions and experiences" (Patton 2002: 348; my italics).

The Frame Reflection Interviews (FRIs)

I was incredibly fortunate that five Australians who lived in Sydney and who had left the DRC as refugees were prepared to spend time being interviewed by me. All five volunteered for semistructured interviews lasting between 45 minutes and an hour. Their ages ranged from 30 to 50 and there were four males and one female. Two were community leaders and approached by me via email. Two volunteered after I had spoken about the research at a community gathering—a church service in western Sydney—and one after learning of the research through a settlement services program the interviewee worked for. The interviewees chose the locations of the interviews, and they took place in homes, workplaces, and an empty public park.

Although the number of people commenting was not extensive, the point of these commentaries was to stimulate my own frame reflection and this most certainly took place. Through their responses they shifted the kaleidoscope so that I came to see situations in new ways. It is worth noting that from the very first moment of this process the scope of my thinking changed. Interviewees self-selected; my original parameters for interlocutors were any former refugee from the DRC, but my interviewees showed me that was wrong. People excluded themselves if they were not from Goma specifically. This was my first big pointer to the fact my research was not about representation alone. Something more complex and important was occurring in Goma, it was a place that needed to be understood on its own terms.

Not right then, but eventually, this led me to the understanding of Goma as a borderland.

The first interview, which was with a community leader and journalist, was more exploratory and less structured than the following four. However, similar themes were covered in all five interviews. These themes can be broadly broken up into the following four categories:

- Representations of Africa
- Representations of the DRC
- Representations of refugees
- Discussions around the practice of foreign correspondence

In the final four interviews I showed the interviewees three two-minute video clips of reports by western foreign correspondents from sub-Saharan African refugee camps and asked them to comment on the journalist's reporting (questions in appendix A). Although I utilized set questions, these semistructured interviews were designed to maximize opportunities for the interviewees to develop their own particular concerns, and for me to follow up on them. This was crucial to the success of the interviews as a process of gathering feedback on alternative frames of interpretation.

The interviews were *not* designed to try and discover a definitive discourse among former refugees from the east of the DRC now living in Sydney—that could not have been achieved with such a small sample. Instead, they were designed to add depth and diversity to the *individual journalist's* (my) primary frameworks of interpretation. The interviewees confirmed, expanded, and challenged the research engaged in up to that point.

Confirmation and Expansion: Feeling the Effects of the Africanist Discourse

One of the most significant results of the FRIs was confirmation of the fact that all the interviewees had concerns similar to those expressed in the literature in terms of the Africanist discourse. This means that the interviews can work as a research tool for journalists looking to report in postcolonial contexts. Instead of relying on books, they can gain a powerful and deep understanding direct from interlocutors. Analyzing the four interviews in NVivo, I found 23 instances of discussion relevant to Africanist discourse. There was broad discussion of colonial-minded reporting, the language chosen by reporters, an overemphasis on poverty as representative of all African experiences, and the false impressions created by foreign correspondence. Moreover, all interviewees were concerned in terms of the overly negative portrayal of Africa that results in a distorted or partial picture of the continent. These concerns weren't always expressed in direct response to questions. For example, during one interview I was invited to share the family meal and I was told: "Don't believe what you hear, we have plenty of food in my country." Specific tropes were also brought up: two interviewees mentioned the overemphasis on African animals as the key residents of the continent (animal trope), and the most media-literate interviewee referred to the big-man trope, specifically using the words "big man." This interviewee also referred to the poverty porn image of the African refugee. This is the transcript regarding the latter:

CG: In terms of the images coming out, do you think they represent you and your experience?

GRAHAM: [2]Not always. Because what tends to get through is the file images, it's just a constant level of recycling, it's the same old images of suffering and kids with no clothes, especially in the context of the African refugee.

Speaking from an African point of view, we always see the kids with flies settling around their faces, mothers, malnourished and with kids on their backs or hanging onto the breasts when feeding, and that's all you see.

In terms of recommending the FRIs as a new research practice for foreign correspondence, what can be seen here is this technique's potential to act as an excellent shortcut for knowledge about problematic discourses and framings. This knowledge is given extra empathic transfer through the fact that the interviewees linked these representational concerns to their treatment in Australia. This was true in terms of being devalued both because they were Africans and because they were refugees, with one or both of those concerns brought up by *every* interviewee.

CG: I want you to think specifically about when you see refugees from sub-Saharan Africa in the media. Do you have any feelings or opinions about the way refugees are shown?

MARIA: When show in refugee camps show just vulnerable people, that's what I see. They show them in a very bad position and that's what they are going through, even worse. But some refugee camps sometimes the women are educated, they do some activities. Maybe they should show stuff like that too, to see that those people are not just sitting there.

In this exchange, Maria seemed to be answering the query I had already developed about whether we should be showing more of the entrepreneurial activity taking place in refugee camps. Later, she also linked this to the way people who have been refugees are seen in broad Australian public discourse:

CG: Do you think the media depiction of refugee camps affects your life here in Australia?

MARIA: I think it's individual. When you come here you have your own plan with your time what you are going to do here.

CG: The way that Australians respond to you, do you think the media has any impact on them?

MARIA: Yeah, because they show refugees is just as people that can do nothing, are just like lazy, and that annoys me because these women they have to wake up in the morning and go fetch water, how many kilometers, they have to go to the farm. How would you call people like that lazy?

Similarly Graham said:

> You never see the other side of life in the refugee camps where there are
> people that are trying to make the most of what they have there. You don't
> see the people who are trying to subsidize whatever rations they receive by
> growing little crops by the side of their tents. You don't see that. You don't
> see the support groups of women who are banding together to support
> other women facing all manner of abuses in the camp. You don't see the
> groups, the young people coming together to create theater groups, social
> groups to support each other. You don't see the male groups, men coming
> together to discuss strategies for issues affecting them.
>
> So you don't see all that; all you see is the suffering—lack of water, sanita-
> tion, food, and other amenities and the conflicts between the refugee com-
> munity and the host community. But there's another side to it and this now
> presents an issue to locals in Australia. When they hear another refugee
> coming in, all they think is the negatives they've been seeing; they don't see
> the contributions these people might bring, what sort of skills and what sort
> of knowledge these guys might bring with them, because they have those
> skills but unfortunately it hasn't been shared with us.

Aristotle said:

> At TAFE[3] people ask how I came to Australia. When I tell them, I lose some
> value. . . . Some people are interested in my story, but others are "Refugees
> useless people."

At this point the interviewees were confirming the existence of issues around
victim representation and lack of agency, and expanding on the importance
of this issue by connecting the effect of these agency-less representations
to their life in Australia. However, it should be noted that the interviewees'
comments were directed not only at the depiction of African refugee camps,
but also at the toxic debate in the media and in the polity regarding refugee
arrivals.

Speaking more broadly in regards to representations of Africa, Thomas
felt Australian media should "sometimes show what is happening in capital
cities in Africa because most people in Australia think that in Africa there is
nothing, just forest and lion and whatever." Even his own children believed
that. This became apparent to him when he arrived home after a trip to the

DRC with high-quality sports clothes. His children asked him how he had found time to *make those clothes*, believing that there was no way they could have been manufactured and bought in the DRC! Graham succinctly put it thus: "I think Africa is still seen as a basket case and Africans are perhaps viewed in the same way."

In this respect, and although this sample is too small to draw conclusions, the FRIs replicated, in Australia, the finding in an English study that representations of Africa as poor and developing affect how people of African heritage are viewed in England (Coulter 1989).

Further, the interviewees expanded upon the issue of Africa being devalued by bringing up an aspect that hadn't been raised in the literature. Two interviewees spoke about how you can see the devaluing of Africa by comparing the quality of African refugee camps shown in the clips to, in particular, given the news at the time, refugee camps for Syrians. These two interviewees noted the disparity in the quality of the shelter provided, with Ben calling it outright "discrimination." He said: "If you go to Syria now, there is no house built with mud. I think refugee have same meaning in Syria, Iraq, Kakuma, so have to live in the same condition. I can say that it's a kind of discrimination." Similarly, Thomas said: "If I see Syrian refugees, compared to Africa, you can see well-organized [relief] compared to Africa. That is a world injustice. All refugees should be treated equally." I was particularly struck here by the value of the FRIs, considering that I had been looking at the same images and had not noticed the difference. Also obvious in this conversation was the global power dynamics that impact at the most local of levels.

Later, I would realize this conversation was the starting point to my eventual understanding of the local-global dynamic inherent in borderlands.

The FRIs confirmed the importance of respecting geographic specificity and not allowing the vagueness of the dark-continent trope to reign. As mentioned, this was revealed from the very outset when several former refugees initially showed interest in being interviewed but decided that they were not appropriate for the research after they discovered I would only be going to Goma. This was the case even for those who were from a town only a few hours away. Also, at various times in the FRIs, three different interviewees noted the importance of understanding local languages and getting facts regarding culture and geography correct so as to produce the best reports. All the interviewees, either implicitly or explicitly, noted the importance of being specific in reporting and avoiding generalizations. This specificity was connected to respect—respecting the diversity of the DRC and respecting the need to understand the complex situation people in conflict zones find themselves in. For example, Ben noted that most foreign journalists reporting from Goma focus on the issue of sexual violence against women, but there are many more issues to report on in Goma. What the FRIs helped highlight is that avoiding generalizations is also an engagement with the complexity of modern, long-running crisis zones.

Specificity for the interviewees also meant acknowledging the uniqueness of each person in the refugee camp, with four out of the five interviewees discussing the need to ask questions about people's lives before they were forced to flee, and their hopes for the future. As Graham said: "Anything to bring to light the human side of the experience is worthwhile, because journalists and [media] consumers have to remember human beings. We're seen as numbers, statistics, rather than actual beings with dreams and hopes."

Knowledge beyond the Africanist Discourse

In my NVivo analysis of the FRI transcripts 20–25 percent of each interview is categorized as "my learnings," by which I mean completely new information that I had not come across in any of my reading. For me this proves the value of FRIs as dynamic process of learning with interlocutors. It is also a further justification for modeling the FRI on the basic building block of journalism research—background reading followed by the interview. However, much of the information I was hearing was so new that I failed to grasp the full knowledge until much later—in fact, not until I had come back from Goma and started to reflexively analyze my experiences. For example, three of my interlocutors talked about the mistrust I would encounter as an outsider. Aristotle put it most bluntly, stating: "People where we live know that wars start in western countries and are finished in Africa, so if you go there you will not get enough information. If any person goes there, we are not happy with western nations. That is why you can't get enough information." Maria linked this distrust to people coming into camps, gathering information, and then disappearing and not knowing what might be the result of this news, while Graham highlighted that in the modern age it is easy for camp authorities "both at NGO and general security level" to find the report, so any critical comments will have repercussions, "and it could be the difference between you leaving that camp or not." Graham related that he knew of a case where someone had "talked themselves out of resettlement." He said: "In the long run there is a level of mistrust not only of reporters but of authorities based on how the authorities treat people." It was only after experiencing the social geography of Goma for myself, and then integrating this with critical literature around the way the space of international intervention works, that I was able to understand how this new knowledge might be integrated into reporting, and this is what is discussed in chapter 5.

The FRIs provided new knowledge and direct contradiction of the literature; this latter point was a challenge I needed to deal with immediately. The research plan I had constructed up to the point of the FRIs was most tested through the responses to the three videos clips shown as part of the situated frame reflection. The second video, set in the Kenyan refugee camp Dadaab, was the one that followed most closely the framing of the stereotypical victim story of refugees in sub-Saharan Africa. However, this was the video preferred by the interviewees—*despite the fact* that they had *all* also expressed concerns regarding stereotypical images of Africa and/or African refugees.

In analyzing the responses I concluded that the interviewees felt the refugees in the second video were in the worst situation possible. Ben, for example, said that "if you are talking about refugees, these are the refugees," and that they required immediate assistance, which is what the journalist was advocating through a transactional frame. This element of advocating was appreciated by the former refugees. For example, Maria said: "I preferred the second one. It elaborates more about the issues. It was kind of on their side, presenting to the world what these people are going through. It was not like an interrogation." Maria contrasted this sort of advocating to the explicit poverty porn-style reporting of DRC rape victims. She equated the intrusive details desired by the journalists with a need to "prove" the rapes are happening. She said: "I don't think those women are lying for journalists to have to go get the truth to say these are the persons. No, I don't think anyone can lie about something like that. It's happening, it's true. Just collect the information and don't expose women on TV." Maria further pointed out that this sort of exposure was the opposite of support for the women: "You have to think about what will happen after you [the foreign correspondent] go home to your place. That woman will be in trouble because when they say it on TV, even entire family knows now, the little children on the street, everywhere. They lose their dignity in the community." Thomas also referred to the second video as advocating for the refugees and linked this to the transactional frame: "The journalist is trying to report what is happening so as to touch the heart of some people, including some institutions, to provide assistance to those refugees, which is good thing."

The apparent contradiction in the preference for the second video, despite the interviewees also wishing to move representations away from victim images, could thus be reconciled with the importance placed on the role of journalism in getting information out. Indeed, all of the interviewees referred to the importance of journalists as conduits of information and, through their answers, showed that they valued this role. This aligns with journalists' conceptions of themselves as "bearing witness," a concept that this research will explore in more detail in later chapters. However, even with this reconciling of the two perspectives via the idea of "getting information out" (what Schön and Rein would call "reframing through co-design") the FRIs proved a challenge to my critique regarding the transactional frame.

Another challenge to the research arose in the idea of reporting complexity. All five of my frame reflection interviewees believed reporting complexity was important, but all five discussed barriers making it difficult or impossible to achieve this. The barriers cited included the following:

- Lack of proper research before leaving
- Lack of good-quality support for the foreign correspondents from knowledgeable locals once there
- Lack of time to investigate before filing stories
- Self-censorship due to the need to reaccess sites like refugee camps that are controlled by government and nongovernment authorities
- Powerful political and economic interests making investigative reporting highly dangerous

Early commenters on this research were surprised at the media literacy of my interviewees in terms of their knowledge of the processes of journalism. I suggest this media literacy is completely natural for people who have, as refugees, been exposed to overtextualized experiences in their lives. The media are an important part of their lives; it is natural that they will pay more critical attention to its processes than others might. It is also worth remembering that these people volunteered to help in media research, so they were likely to have more than the usual interest in issues of media representation.

In citing the barriers to reportage, again, there is an apparent contradiction here—this wish for investigative reporting doesn't seem to align with Maria's resistance to journalists as interrogators. However, this contradictory dynamic can be again reconciled (or to use Schön and Rein's terminology—reframed) through consideration of power relations. I believe the former refugees wanted to see journalists use the tools of their trade to punch up, not down. Finally, it is worth noting that the barriers cited, particularly the lack of time, self-censorship, and the powerful politico-economic interests, were seen as insurmountable, implying that simplistic or partial reporting is expected even though complex reporting is valued.

Schön and Rein describe the complications that can arise from actors' conflicting frames, and the FRIs certainly revealed such complications between us as actors. In particular all interviewees revealed a strong interest in journalists reporting on the poor conditions in the camps, and on the injustices and arbitrariness of the refugee resettlement system. It was difficult to imagine how, especially in regards to the first issue, this could be done without veering toward the victim-transactional frame I was trying to move away from. The result was the knowledge that I would need to avoid the victim frame without avoiding the facts of the situation.

In focusing such a strong spotlight on conditions in the camps, the FRIs also reinforced a point that Schön and Rein make: all communication takes place in space and this has an effect. They elaborate on the point thus:

> Because there are no institutional vacuums, interpersonal discourse must have an institutional locus within some larger social system. Even a chat between close friends occurs in the institutional setting of someone's house or a walk around the park. This institutional embedding is important to the nature of discourse in several ways. The institutional context may carry its own characteristic perspectives and ways of framing issues, or it may offer particular roles, channels, and norms for discussion and debate. (1994: 31)

I was reminded that I would be working inside a refugee camp and it is correct that people's concerns would be influenced by that institutional context. This is why I paid attention to people's description of their hunger and made it the opening focus of one of my feature articles. I doubt I would have done this if not for the FRIs.

Taking Stock of the FRIs

In terms of experimentation I came away from the FRIs considering them a success. I note that in some ways they may be considered simply part of normal journalistic practice. Journalists who were likely to travel to the DRC to report and had been given enough forewarning might try make contacts in the DRC community in Australia and, over time, would be exposed to a variety of viewpoints.

That local immigrant communities should be accessed by reporters before heading overseas has already been advocated by scholars looking for solutions to the pitfalls of current foreign correspondence (Erickson and Hamilton 2006: 43–44). I consider the FRIs as furthering this argument; they worked as an excellent starting point and shortcut, allowing deep reflection on frames to take place quickly—reflection that could lead to completely new storylines being developed. In this respect they could also be seen as a focused form of "small talk." Foreign correspondent Jim Lederman states that hours of small talk with nonelites is needed to break the "tunnel vision" of dominant storylines, and that small talk will usually not take place (1992: 25).

The FRIs, particularly in the context of areas of journalism that have been highly criticized, like foreign correspondents' reporting from sub-Saharan Africa, could also be seen as a conscious educational tool aimed at "coming in right." This concept of "coming in right" is raised by former National Public Radio and CNN correspondent and bureau chief in Johannesburg Charlayne Hunter-Gault. Hunter-Gault said her experiences as an African American woman, and her early experiences as a reporter specializing in the urban African American community, "propelled [her] to go beyond traditional sources" and helped her resist the urge to thematization and stereotyping as a foreign correspondent in Africa (2006: 109–12). FRIs, used consciously as part of our practice, could be a more conscious, systematic attempt to ensure we "come in right."

FRIs could also be considered to have a role within the global news ecology outlined by Cottle: "Through the circulating flows and communications cross-traffic of today's news ecology, an expanded array of voices—global-local-West-rest, elite-ordinary, expert-lay, military-civilian—can now sometimes enter the frame and challenge the parameters and preferred terms of public discourse." However, Cottle adds that these voices do still have "differing degrees of access and possibilities of success" (2010: 482). Using FRIs, individual journalists can seek to consciously diversify their frames as a way of tilting (although not evening up) the "degree of access" and "possibilities of success" for those who do not always have the power to challenge the "preferred terms" of public discourse. In an unequal global society, fairness is a journalistic ideal that needs to be constantly worked at and redefined. One way of fostering that ideal is to aim for a diversity of frames speaking to a diversity of discourses.

Having said so much in favor of the FRIs, I emphasize that I was working as a journalism academic, and it would not be easy to introduce this practice to professional newsrooms. Time is a working journalist's most precious resource, and the FRIs took time to design, organize, and conduct. The potential discomfort caused to journalists through exposing themselves to critique in a highly unpredictable and alien environment is another issue. It is one thing to approach someone as a "source," asking the person to provide comment for a news piece you are crafting, in an institutional context in which you are familiar. It is another, much harder, thing to put yourself under strangers' their tutelage and ask them to take you into their world. The first version of simply seeking more diverse sources mirrors the move in recent times to decenter privilege and open "speaking spaces" for nonwhite audiences. As pointed out by Australian journalism academic Tanja Dreher, this may seem like a worthy goal, but in practice it has not always worked out well for the sources involved. Nor does it necessarily shift patterns of power. She points to the example of Muslim women in Australia "who have become highly visible in public debate during the 'war on terror' but have also found it extremely difficult to shift news agendas and to be heard on their own terms, instead being asked to constantly respond to the concerns and stereotypes of 'mainstream' audiences" (Dreher 2009: 4).

Dreher, building on the work of feminist scholar Krista Ratcliffe, argues that the more ethical position is "eavesdropping with permission" (2009), that one should seek to gain access to discourses in a way that redistributes risk and discomfort toward the researcher. Taking the concept back to its lexical roots of standing under the eaves in an attempt to gain knowledge (possibly in the uncomfortable position that includes getting wet!) this "choosing to stand outside . . . in an uncomfortable spot . . . on the border knowing and not knowing . . . granting others the inside position . . . listening to learn" (Ratcliffe in Dreher 2009: 12–13) becomes a tactic for receptivity to the discourses of others. Eavesdropping with permission took place for me inside the church hall, and some discomfort comes with decentering your power, asking to put yourself in another person's space and simply listen.

There is also a decentering of power that comes with asking laypeople what they think you, the professional, should be reporting. There is discomfort that makes the FRIs both hard and humbling. This is in fact another reason to engage in the FRIs but could also make them difficult to take on, especially if operating without institutional support.

In emphasizing the discomfort and risk placed on the researcher the FRIs can be distinguished from attempts in the 1990s of news organizations to engage in public journalism. Although these attempts were aimed at helping audiences gain more control over news narratives, they have been criticized as in fact being highly top-down endeavors that engaged mainly with already privileged audience demographics (Min 2018: 30–37).

While the FRIs are a one-on-one process, in theory, easy for any journalist to undertake as part of their professionalism, in practice the FRIs may require some sort of partnership arrangement between academia and media organizations. This can help with the time constraint, as the academic researchers can be the go-between to find the community participants. Area experts from within the academic community can also be drawn upon to help craft the questions of the FRIs for each specific situation. This partnership model is how the FRIs have worked so far in two different situations. In the first instance, working in my role at the Centre for Media Transition, at the University of Technology Sydney, and joining forces with an area expert, I was able to design a new set of FRI questions for a correspondent about to be positioned in Jakarta, Indonesia (Giotis and Hall 2021).

The second trial involved four journalists from Australia's national broadcaster, the Australian Broadcasting Corporation (ABC). Two were members of the Asia Pacific newsroom; their remit included reporting on Muslim communities, both in Australia and in the Asia Pacific. Two were not specifically tasked with overseas reporting but had a keen interest in issues of diversity. The project centered around tackling Islamophobia. It proved the value of the FRIs as a *glocal* practice beyond foreign correspondence. In this research the issues of local representation were tied with international discourses of Muslim communities, and Muslim refugees in particular. The interviewees explained to the journalists how the representations produced by foreign correspondence impacted the lives of young Muslim Australians (Giotis 2021).

The agency of the journalists in both projects must be acknowledged, as all volunteered to take part in the FRIs on top of their regular duties. Again, time pressure is a barrier to the take-up of the FRIs. However, the FRIs can also be seen as a structure aiding agency. One of the outcomes reported by journalists in both trials was a greater determination to complexify and diversify reports (Giotis 2021; Giotis and Hall 2021). Moreover, this determination can lead to projects that gain institutional support. In the case of the ABC project, one of the journalists and one of the interviewees got together after the FRI process and, after gaining the support of a key news editor, created an Iftar[4] event bringing together 20 ABC journalists with 20 members of the Australian Muslim community at the Islamic Sciences and Research Academy of Australia. The event included sets of questions distributed among the tables, a "talk show" panel, and a pitching news stories activity; three of the stories pitched by the community members were commissioned on the spot by ABC editors.

Writing the Sauti Article

I set off to meet the women of the Sauti group having already completed my postcolonial analysis of the Africanist discourse frames and my FRIs—I should have been in a good position. But this thinking about frames would not be the only information from Australia structuring the experience. As noted earlier, all "interpersonal discourse must have an institutional locus within some larger social system" (Schön and Rein 1994: 31). In the case of the Sauti article that larger system was that of INGOs. I first heard about the women of Sauti at a conference in Australia and I was urged to get in touch with them by an INGO worker who, like me, was concerned with shifting the one-dimensional victim image of women in conflict zones in sub-Saharan Africa. In some ways the story design had already been set before I left Australia. Compounding this initial setting up, the interview itself took place in the offices of the INGO that first helped coordinate the group—the conversation was bound to be centered on the organizational structure. The problem was compounded again by the fact this was my first time engaging in journalism that was also academic research, and I had to adhere to the university rules of explaining, at length, the purpose of the research. In short, the context surrounding the creation of the Sauti article was overstructured by institutional concerns. I do not consider the interview to have been successful as a process of discovering new knowledge, and I do not think that I was able to enter into the world of my interviewees and see it through their frameworks of interpretation.

This lackluster performance in the interview can be seen in the writing . . .

Dateline: Goma, DRC, August 26, 2014

What follows is an exact reproduction of my draft article. As part of the research experiment I captured my thinking, as I was writing, as notes in italics, and instead of deleting sentences I used the strikethrough function. These notes (sans typos) are exactly what was thought during the time of writing in my hotel room in Goma.

How will I get complexity in? MSF workers story? "rebels are from our families."

What about avoiding tropes? Well, it's quite possible that I can't, but perhaps talking about very western idea of protest, banner, and placards will at least help.

One by one the women of Sauti Ya Akina Mama Mkongomani arrive. These five members of the organization which translates to The Voice of Congolese Women have busy lives, they have jobs, families, they have key positions in grassroots civil society organizations and they were only told about the interview yesterday afternoon.

~~They have come because these women believe in getting their voices heard and they will fight to make sure that they are. As proof of the lengths to which they will go I'm shown a picture of the group taken at the time of the XX peace negotiations in Kinshasa. Kinshasa is the capital of the Democratic Republic of Congo~~ *Taking too long to get to the women. Too impersonal—but then I do like the intro.*

Yet they have come this morning because they believe in being heard, and of being heard together.

Rachel Kembe is the first to arrive. She is known as Mama Rachel and has a severe injury from an attack during one of the earlier phases of the protracted war here in the east of the Democratic Republic of Congo. She uses a caliper and when she walks her hip seems to be disjointed from her leg, yet it is Rachel who reaches up to a high nail on the wall and pulls off a calendar, she flips through the months until she finds the picture she wants. It is the Sauti Ya Akina Mama Mkongomani group at what's obviously a protest rally, they have a large banner and poster placards. Each placard is different and each represents a policy position for women.

~~which they believe should have been part of the M23 peace negotiations that took place last year.~~ *Confusing for readers without context and history. This is so hard to write! Also realize I should have asked more of their histories.*

Their experiences in the war. But then afraid of causing trauma. ~~. The event is in Kinshasa, the DRC capital~~

Mama Rachel explains that this picture was taken last year in the DRC capital Kinshasa during the national dialogue which followed what's known as the M23 rebellion, one of the most recent, dramatic chapters in the DRC's war in the east, when rebels overran much of the eastern region of North and South Kivu causing 800,000 to flee the fighting between them and government troops. M23 also captured the largest Eastern city, Goma. ~~and causing the large international community to flee~~ *so what? What about the locals.*—the city in which we're sitting and talking now.

The distance between Goma in the east and the capital Kinshasa is huge, 1,500 km to be exact, planes are the only reliable and realistic form of transport and they aren't cheap but the Sauti group was determined to have the women of Goma represented. When they found they couldn't gain access to the meeting they wrote up the placards and started demonstrating outside in the street so that they couldn't be ignored. In the end the strategy worked and on their return from Kinshasa the governor of North Kivu province asked for a meeting with them to hear their concerns.

I'm shown the picture and then other members of the group want to have a look too. Although these are five key members who have been with the group since its formation seven years ago not all were there in Kinshasa. The Sauti group is actually quite large. It has 100 members just in Goma, some have joined up as individuals but around half represent a grassroots women's organization. Initially the group started with 30 members, all of them representatives of an organization, the idea being to create a strong peak body that could get women's issues on the political agenda.

Sitting at the table next to Mama Rachel is 49-year-old Faulestine Matsindu Butaintinwa, *I initially had her refugee background here in her initial introduction but didn't want her to be defined by her "refugee-ness"* the organization she represents is called Fupros and is a grassroots women's collective that tries to help vulnerable women as well as more general awareness raising against gender discrimination. *I should have asked the question, what sort of discrimination.* ~~fights against the discrimination women face~~. A refugee who arrived in Goma 20 years ago "at the beginning of the war" *(from Rwanda? Just not sure. Felt constrained asking about those details—felt not essential to the story but that's stupid—of course it's essential, and essential to my research, this is still about the representation of refugees!)* she tells me she felt compelled to play a role in Fupros because she knew how to write and that gave her a valuable skill. ~~in 1994 and has made this city her home for the 20 years since. since a Goma resident for 20 years she arrived as a refugee in 1994 at the beginning of the~~ *really struggling with this, I just don't know enough about her history and that's exactly what I was supposed to ask about! I could have asked more, not just about her victimhood but started further back.*

On Mama Rachel's other side is 24-year-old Justine Kitumaini, who was just 19 when she joined the group as one of the founding members, and is spokesperson for an organization that helps orphans, single mothers and prostitutes ~~(a growing industry both thanks to the war and the supply of large groups of international workers).~~ *I can't say this. I don't know it is* GROWING. The latter is an industry which traditionally thrives during war and that is true here too especially when women's livelihoods are disrupted and when they have been raped and so struggle to return to the community. What's more in Goma prostitutes also find customers in the international community posted here as part of the UN mandate to keep the peace. *OK that's true. Feels taboo to say it! But I saw it with my own eyes at Coco Jumbo. Why does it feel taboo? Feels like I'm going against the orthodoxy of the Good Samaritan. Also, shouldn't make it out like an imported problem, there is also the issue of war-rape.* Miss Kitumaini's organization provides basic support like food and shelter but also fights to raise awareness about HIV prevention.

Marie-Claire Masika, 54, former resident of Rutshuru, who was forced to flee because of the war, works with an organization that supports women in agriculture, <u>an industry which should thrive in the fertile climate but is often either disrupted by fighting, with women in particular vulnerable to attack, so that the crops fail or taken over by the armed groups who demand high percentages for protection.</u> *Not sure about this, again feeling the tension between complexity and chaos. In the end also disrupting the flow of the story about these women. Actually, no it's important—leaving it in.* The final member of the quintet is Collette Kavira, 40, who arrived in Goma in the 1997 war which brought about the end of Zaire and the beginning of the Democratic Republic of Congo. She works for an organization called Federation for Peace in the World, it helps reunite rape victims with their families and also runs education programs for women. Because says Mrs. Kavira, quoting a proverb, "If you educate a woman you educate the whole nation."

The range of the organizations' work reflects the extent of the issues here in what is a hugely complex situation. <u>The 20 years of crisis, which started after the genocide in neighboring Rwanda, and the end of the cold war,</u> *included cold-war because wasn't willing to let it just be tribal politics.* <u>changed the power dynamics of the region,</u> *Do I really have space to put the history in?* has morphed in character and focus but for the people in the Eastern DRC the one constant has been instability caused by armed groups.

The price women have paid is of course well documented but as Sauti Ya Mama Mkongomani demonstrates, they are not just voiceless victims *I don't love this but going to leave it in because so close to title of thesis. —* ~~not that this makes their lives any easier—or the situation less complex.~~ *Telling not showing* The group was recently called upon to try and help with making contact and brokering negotiations with <u>the four? Congolese?</u> Médecins Sans Frontières workers who were kidnapped a year ago (this week, during a military skirmish one, Chantal Kaghoma Vulinzole was able to escape). I ask how they could be expected to find out information and they tell me that in Goma, a <u>city of one million</u> *need to double check this*, which has grown four-fold in the last 20 years, largely through refugees and internally displaced people, information (and rumor) always can be found.

<u>"Those rebels, even if they're there now, they were born in our families, that's the reason we can talk to them,"</u> says Mrs. <u>Butaintinwa</u>. *(I thought this would help get across the complexity—has it? Or just made seem chaotic?)* Of course, she doesn't mean their literal families but the fact is even though Goma may have the appearance of calm and a normal city (albeit with a lot of UN and military around) the community is constantly interwoven with the conflict as it steals their sons and daughters.

And that is why the Sauti Ya Mama Mkongomani group, representing 50 community organizations is so determined to have their voices heard— even if it means sometimes shocking the international NGO that helped the group form.

Action Aid supported the set up of Sauti Ya Mama Mkongomani based on its successful experience of bringing together women's organizations to create a peak lobbying group in Kenya. However, the group surprised its sponsors when the 14th Ordinary East Africa Heads of State Summit decided to put the DRC crisis on its agenda. Sauti Ya Mama Mkongomani decided they should go to Nairobi and put their case in person. There wasn't budget to support them but the women went on a fundraising drive and booked their flights. When they arrived in Kenya Action Aid realized the force it had to reckon with and provided their accommodation.

The Goma group has also inspired organizations from four other areas of the DRC to form a peak group under the banner of Sauti Ya Mama Mkongomani in their region. However, if the group is going to meet their rather ambitious goal of being known around the world it needs to raise profile and build capacity, a task that is especially important now that the funding from Action Aid has stopped (although there is still strong support and links to the organization—for example we are meeting in the Action Aid Goma office).

~~From me they ask only transport money~~ *Transactional frame—but then again that's what they want! Desperate for training especially in new technologies and social media.*

Mrs. Kavira tells me they have a long list of training they want to undertake and at the top of the list is social media, both for awareness raising and information monitoring.

Peace one day concert—news hook??? but then also falls into bullshit ray of sunshine at the end frame—struggling here . . . frame of struggling?

It's going to be one more struggle for these women to take on in a difficult environment—but at least this struggle is very much of their own choice.

Hmmm, very aware of critique of ray of sunshine but also very aware of journalistic genre. Would be so easy to write something like "a sensible suggestion in a senseless world" but that's claptrap. Could go straight to Action Aid but then becomes very transactional. Last sentence so important for overall frame created. Decided on struggle as most representative of the situation.

Like more information or to support contact Action Aid—perhaps Action Aid Australia's number?

Peace One Day concert website link. *But does this make it INGO-y again? Or does this make it cosmopolitan?*

Reflection in and on Writing and More Co-design Work

Journalists are supposed to use frames for reasons of efficiency: "Media or news frames serve as working routines for journalists, allowing them to quickly identify and classify information and 'to package it for efficient relay to their audiences'" (Gitlin in Scheufele 2000: 306). In practicing frame reflection while writing up the Sauti piece, I was practicing the opposite of efficiency. As my thinking notes reveal I found myself struggling sentence by sentence as I tried to avoid frames related to tropes of the Africanist discourse. For example, in discussing their work, the women talked about their role in negotiating with rebels for the release of hostages:

> "Those rebels, even if they're there now, they were born in our families, that's the reason we can talk to them," says Mrs. Butaintinwa. *(I thought this would help get across the complexity—has it? Or just made seem chaotic?)*

Or, in another example, I agonized about whether to include the upcoming Peace One Day concert in the story. The concert provided a timely news hook to be reporting from Goma at this time. However, I was worried that it would activate the transactional frame. The fact that it would be placed as a link at the very end of the article made its potential impact on the overall frame extremely strong—a *last*-ing impression. Similarly, as can be seen in the draft, I agonized over the final paragraph of the article.

In the end I accepted the fact that it is probably impossible to avoid Africanist discourse altogether, especially when one takes into account the different ways each individual reader will decode the article. As my notes at the time tell me, I hoped that opening with the *"very western idea of protest, banner and placards* [would] *at least help"* steer readers away from the ideas of the Africanist discourse.

I had expected this article to take longer than a normal article to write, but I was surprised at how much longer and at how much the frame reflection process interfered with the process of writing the story. Journalism is a time-sensitive profession—I would need to find another way of practicing frame reflection while writing before attempting to introduce this technique to the profession.

The representational choices were difficult but they were made. The next step in the experiment was to introduce elements of co-design by again involving multiple actors—this time by sending the draft article (sans thinking notes of course!) to people from a variety of concerned sectors and asking for feedback.

The actors I identified for co-design were two development experts, a former *Foreign Correspondent* producer, a journalist-academic, one of my FRI respondents who is also a journalist, and a high school history teacher who is also my friend. I sent each of them a list of questions via email and asked for feedback before heading to Mugunga camp. However, the former foreign correspondent did not respond at all and the journalist-academic did not reply in time for the visit to Mugunga and also did not reply in the questionnaire format but via a Skype call. It should also be noted that I had not intended to get feedback from two development experts; however, the second expert was from Action Aid Australia, and had put me in touch with the Sauti group in the first place and asked to see the article, so I took the opportunity to get her feedback too. Her relationship to the group is highly relevant to her responses.

The questions attempted to examine whether the framing seemed familiar or new to the reader and further asked specifically about the concerns I had in terms of complexity versus chaos and the victim framing. The rationale for each question and the answers in full and exactly as received are presented in appendix B.

Overall, the email responses were complimentary. In terms of my concern as to whether I was conveying complexity or chaos, all respondents answered that the story conveyed both, but most said the complexity frame was stronger. Only my FRI respondent said that both frames were equally strong and none said that the chaos frame was strongest. I felt this was a good result. I was also happy that people were engaged by the story and felt they had gained new knowledge. I was struck, but not surprised, with how each response revealed both the institutional frameworks of responders and their own personal interests. For example, the journalist discussed the need for a more balanced picture, the historian highlighted the need for more background information on the war, and the development expert was interested in the fact that the group was advocacy focused.

In the unstructured Skype conversation with the journalist-academic two key points of feedback came through. The first concern, very much of the journalism profession, related to how the article reads. According to this feedback, the introduction was not arresting enough to grab readers; the inclusion of the full title of the group, Sauti Ya Akina Mama Mkongomani, would "lose" readers; there was no need to insert myself into the story; it was not clear how the interview had taken place (i.e., through a translator); there was more telling than showing; far more detail was needed to make the story more believable; and it was written too much as if I was employed by Action Aid.

The second area of concern was that the women were not showcased enough. The journalist commented that they "hadn't flowered under my gaze" and that I hadn't constructed a story of "survivorship" because I hadn't constructed a heroine. In particular, I was urged to introduce the story with Rachel Kembe's injury, and when I expressed concern that this was speaking to the victim frame, I was given this advice: "I honestly don't get a feeling of what these women are like and that is more colonising than victimhood."

At this point the experiment with co-design took a dramatic turn. Although all the feedback on the draft was valuable and valued, the strong critique from the journalist-academic, which mirrored my own concerns with the article from the point of view of a journalist, rather than a researcher, meant that the journalistic institutional frame of "readability" was most strongly present in the rewrite. Indeed, until receiving the feedback from the journalist-academic I had not considered doing a full rewrite of the Sauti article. One of the key forces acting on journalists is their habitus, and the structure of my habitus up to that point meant that the response from the journalist-academic would carry most weight.

As discussed in the introduction, habitus is a concept championed by sociologist Pierre Bourdieu as "the strategy generating principle enabling agents to cope with unforeseen and ever-changing situations . . . a system of lasting, transposable dispositions which, integrating past experiences, functions at every moment [variable by age, time, and place] as a matrix of perceptions, appreciations and actions" (Garnham and Williams 1980: 212–13). Because habitus always acts in relation to past experiences, primary socialization processes are crucial. However, as you move between fields your habitus moves with you, and to operate successfully in the fields of your endeavor, your habitus must take on the "logic of the field." In this sense it is both structural and personal. My socialization as a journalist, my habitus, is absolutely crucial to everything that happened in Goma—in a sense I carried my western newsrooms with me, inside me. Habitus is appropriate as a conceptual tool here because it emphasizes both agency and structure. Every decision I have taken in my career as a journalist has been my own, but every decision has also taken place within the structural constraints of the particular, *current* rules of the game.

Schön and Rein argue institutional frameworks are key guides in the decision-making process, calling these institutional action frames (1994: 33–34). As a journalist, I was socialized into a system, and I am *used to* referring my work up a seniority chain. That inevitably resulted (especially in my early socialization) in changes that were justified by those up the chain via the institutional values of journalism, especially the value of "readability." I would then accept those changes (and implicitly those values), thus entrenching this *disposition* as part of my habitus. I wished to modify my institutional action frames via the FRIs; however, alternative values and practices had obviously *not yet* become an enduring part of my habitus. I had no internal structural frame able to override the value of "readability" as all important.

I wish to be clear here. The critiques by the journalist-academic were valid, but my easy capitulation to the institutional action frame of "readability" is problematic. This is because that one word acts as irrefutable justification for a constellation of professional decision-making processes. This includes one of the key hierarchical structures of our profession—news values. Conley and Lamble describe standard news value criteria as "impact; conflict; time-liness; proximity; prominence; currency; human interest; the unusual" (2006: 83). While news values may seem neutral when presented as "parsi-monious lists" (Waisbord 2013: 183), in truth these values intertwine with less-than-ideal reportage in foreign correspondence. For example, in exam-ining the coverage, in Norwegian newspapers, of Zimbabwe's 2000 consti-tutional crisis Ndlela (2005) posits that the news value of negativity turned "the complex Zimbabwean issue into a 'typical' African story of tragedy and despair" (2005: 89). On the news value of "proximity" Ndlela's analysis found that stories focused on the white farmers, whom westerners could "identify with," as opposed to the major victims of the violence, the black laborers who did not support Mugabe politically. Thus, the "proximity" news value con-tributed to downplaying the politics and presenting the issue in racial terms (Ndlela 2005: 83).

A key reason news values are problematic is because they prioritize existing definitions of the world. This has been known since 1965, when Galtung and Ruge drew up the first list of news values. This was a seminal event in media studies, but a byproduct of their main research question trying to identify ways that international reporting can be improved. The improvement that they, and others since, have suggested is that journalists should try to con-sciously counteract prevailing news values (O'Neill and Harcup 2008: 170–71). The difficulty, however, in putting that advice into action is that this requires going against existing institutional action frames. In a classic 1980 Australian study journalists were first asked to describe what makes a story newsworthy, and they came up with various "imprecise, rambling and vague" answers (Baker 1980: 140). Those same journalists were then asked to rank stories by newsworthiness, and instead of producing a variety of rankings, corresponding to the variety of their original answers concerning what news is, they instead came up with almost identical rankings. This result led Baker to conclude "news is not so much what Newsmen [*sic*] make it as what they know the News Organization they work for sees as News," and this relies on a "socialization process localized within their own news organization, but also loosely linked through similarities to a broader Australian-wide journalistic institution" (Baker 1980: 155). In other words when journalists apply news values they operationalize hegemony.

Despite my knowledge of these problematic aspects I have spent 10 years in newsrooms and so I have been socialized as a journalist. My habitus, my strategy-generating principle for acting in the field, told me that I had to give the newsrooms what was expected.

In rewriting the Sauti article I had to deal with the lack of specific information. This meant redoing interviews to discover more about each organization's work. Also, I needed to learn more about Mama Rachel's injury. The follow-up interviews took place over the phone, with my translator on the phone and with me sitting next to my translator. Not ideal, but I was time poor. The one person I was not able to reach was Mama Rachel. Despite this, through conversations with others, I was able to learn more about her injury, including that it had not been a result of direct conflict but the result of banditry, which was rife around the time of the M23 rebellion. In the interview she had called herself a war victim, and I remembered the other women nodding in agreement. How to incorporate the new information became a major issue. Below is the draft rewrite of the introduction with my thinking notes in italics:

Rachel Kembe is tall, at one time would have been graceful = *victim frame!!*

In December 2011 Rachel Kembe's friends feared for her life. She had been shot in a robbery at her little general store, she wasn't a direct victim of the = *not as emotive and grabbing*

Rachel Kembe is tall and has managed to maintain both grace and forcefulness in her movements despite the fact that she is the survivor of a brutal shooting [*if I leave without details what does it imply?*] which has left her using a caliper and given her a severe limp where her hip looks disjointed from her leg.

As you can see I was oscillating between three points: a commitment to avoiding the victim frame; concerns with the readability/attraction of the story; and the truth of the story. In the end I chose this introduction:

Rachel Kembe is tall and maintains both grace and forcefulness in her movements despite the fact that she is the survivor of a brutal shooting.

Mrs. Kembe, 46, or Mama Rachel, as she's known, lays the blame for the shooting not only at the feet of the bandits who almost killed her, as they robbed her of her small earnings in her general store, three years ago. She also blames the continuing war in the east of the Democratic Republic of Congo (DRC) and the resulting general lawlessness and impunity that too often exists. In fact she calls herself a war victim.

I would never have arrived at this synthesis of ideas if it had not been for the knowledge I gained through the FRIs, and in particular Aristotle's use of the term "blasting war." At one point Aristotle said I must be careful and make sure that the situation remained relatively safe. He said: "There's war but not blasting war like before." His differentiation helped me understand that the constant, ongoing nature of the conflict has underlying effects. The community is still at war, and is still affected by the war, even if there are not large military actions in the immediate vicinity—no "blasting war." As a socially shared category system, "blasting war" helped integrate the truth of the ongoing, volatile nature of modern borderland warfare into my primary frameworks of interpretation.

The rest of the rewritten article[5] largely followed the structure of the first article but with a few more descriptors of the women and their work. For example, I describe Faulestine Matsindu Butaintinwa as "impatient and forthright," a descriptor I felt confident in offering, even if I could not understand what she was saying, as her tone and body language spoke volumes. Normally, I would not do this; I allow interviewees' characters to come through in their quotations. I changed my practice to take into account my language disability. However, it would have been much better to have had access to the interviewees' words because individual words help piece together frames. This issue came up again when I was putting together the Furaha piece and I again had to find a workaround.

Reflecting on Action: Incorporating
FRI Feedback into the Furaha Article

A common theme in the responses from reviewers of the draft story was a wish for further information and details. The one exception was the Action

Aid respondent, who of course had a lot of background information already through knowing these women and their work. This valid feedback corresponded with my own feeling that there was much more I could have learned about their work and their world. The issue was that I had conducted a very long interview with the five women sitting around a table but had received very generalized answers. This, I believe, was the result of the translation process. I also felt I was unable to interject with timely follow-up questions for the same reason. The translation process added a lot of time to the interview, and I could see my interviewees becoming frustrated, especially at the very outset when a lot of time was spent explaining how this was university research and journalism at the same time. This initial frustration made me less willing to add more time onto the interview by pushing for more details.

Before entering Mugunga camp my translator and I discussed streamlining the introduction and that his translations should try to retain more details. I also discussed the fact that I might ask the same question, similarly worded, more than once if I felt the first answer was too vague. This was something I had not done in the Sauti interview. I didn't want to push people to answer something they didn't want to answer, but I also didn't want to assume this was the case before trying a second or third time to find the right question design. Again, not being able to judge people's reactions to the questions myself was a difficulty, and of course reinforces the problem with foreign correspondents not learning local languages. However, there was some improvement in the interview translation process during my time at Mugunga; my driver, guide, and security officer all chipped in and helped as well. For example, one of my interviewees, Chantal Samvura, mentioned that she chose to work in the camp sanitation department because this gave her access to soap so she could keep her family clean. This was obviously important to her and she mimed the action of washing clothes with her hands. This was also consistent with the importance my interviewees in Australia placed in respecting people's attempts to maintain dignity in inhumane conditions. However, my translator did not mention this when relaying Chantal's answer. This omission was picked up by my driver listening in on the interview and he urged my translator to include it. My driver's suggestion seemed to be vocally supported by those listening in, including my interviewee and her friends.

Another opportunity for incorporating feedback came when I saw some young men cutting long planks of wood. Influenced by the fact that one of the Australian respondents had noted the absence of men in the Sauti article,

I interviewed the men and discovered they had started a carpentry business at Mugunga, hoping to support their families.

Sadly, both Samvura and the carpenters were cut out of the version of the Furaha story that was published in the Australian media. These cuts were made due to space constraints. The fact that the two unusual angles to the story, and the two angles that were specifically asked for by my co-designers, were the two elements sacrificed for word count, is perhaps not coincidental. This may in fact be proof of how alien the concerns of subaltern voices are to the mainstream media, and how much work needs to be done in developing a shared, recognizable lexicon for describing borderlands.

Extract from my research diary:

Monday June 30, 2014

OK, so I want to go back to Sunday June 15 when I went to the Congolese church in western Sydney because I do feel like that was a bit of a turning point.

Despite this I came home a bit disheartened with the number of people interested. I don't think I was ready for the diversity of frames in the messiness of life; the fact that people wouldn't be that interested in talking to me. The beauty queen, mother, former refugee, intelligent enough not to want to get involved, not to want to relive her trauma. But as Steve said to me that night, whatever anyone of your friends did today, what you did was more interesting. And that's the point. I'm not trying to claim a definitive list of frames for use in the DRC. I'm simply trying to expose myself to a wider set of discourses, in that sense just being in the band-hall-cum-church was important, feeling responses, spoken and unspoken, to my questions, to my presence, especially at the lunch afterward. Listening to the sermon, on the need to understand your God-given gifts. Understanding, really, for the first time why faith survives despite, or because of, horror; listening to the story of David and Goliath, to the frame it was put in. I think that first exposure helped me to later hear the responses and nuances of my two interviewees in western Sydney after that. Again, traveling out to [X], taking that one-hour plus journey on the train, seeing the library and cost of food. Those things are important.

I used the word "simply" just now, "simply trying to expose myself to a wider set of discourses" but everything is always more difficult than you think it will be and there's a safety in the formula of news. X [former *Foreign Correspondent* producer] made that point too when I met up with him.

I have one last note in my notepad from that weekend. The question, how can I create frames without knowledge? A question sparked by reading books of the DRC's history, both Kwame Nkrumah's and David Van Reybrouck's. The knowledge that Leopold's revenues directly relate to the electronics industry of today. The knowledge that Australians were the first to mine gold in the DRC. These are the bits and bobs of knowledge that frames are built on, these are the nails and rivets essential for the scaffolding to hold.

I have gone back to this research diary extract because, to find a frame to write the next article, I felt I needed to return to the FRIs. I came to this conclusion after first reflecting on the positive and negative attributes of the Sauti article.

As can be seen from the thinking notes in the original write-up of the Sauti article, and from my discussion of the rewrite, I did not find writing while practicing frame reflection easy. I was pleased that the responders to the Sauti article did not evoke the word "victim" when I asked them to use three words to describe the women. I was also pleased that most believed the complexity frame was stronger than the chaos frame. In this respect, the effort taken to focus on frames *while writing* seemed justified. However, the time taken to write while practicing frame reflection remained a concern; it did not seem like something I could recommend to a time-driven profession.

My personal reflection on the rewritten Sauti article led me to conclude that it was not much better than the original. I realized that although the individual elements were better than the original, it lacked what I can only describe as "pizzazz," because I was more focused on portraying agency and complexity than on the overall story frame. In particular, to use Schön and Rein's terminology, it lacked a generative metaphor. This failure was present in both versions, but in the original version the failure was not *absolute*. In the original version, the idea with the intro was to present these women in a way we were used to seeing western women campaigners described: busy, committed, sharing a sense of solidarity—so there was a bit of a metaphor, in the sense of transferring the western notion of strong political grassroots female campaigners to the DRC, and this obviously was successful in some regards, as the historian brought up the idea of being reminded of the suffragettes when reading the article.

It is important to remind ourselves here that there is a difference between metaphors as the tropes of the Africanist discourse that help create a process of sedimentation of certain ideas, and the idea of metaphors as vehicles of movement, metaphors as *metapherein*, that is, transferring the idea of feminist activism from an exclusively western activity to one engaged in by Congolese women. Schön and Rein create this emphasis by taking the word back to its ancient Greek meaning, revealing that its root lies in *movement*: "The familiar and the unfamiliar come to be seen in new ways" (1994: 26–27).[6]

In the second version of Sauti the metaphor of western protest was lost. In fact, in the second version the article suffers from the lack of any cohesive frame, which is why, while the individual elements of the story are improved, the story overall is not. In this analysis I am supported by one of the founders of rhetoric theory, the Greek philosopher Aristotle, who argues that learning is pleasant and that metaphors are the most effective tool for bringing about learning, assuming that the metaphor is not superficial and requires some work to be done on the part of the receiver, and that it is neither too complex nor too unfamiliar for it to be intelligible (Aristotle and Kennedy 2007: 218–19). Because I did not employ a metaphor as *metapherein* for the story, the story lost some of its appeal as a way of learning new information, and this is where the lack of pizzazz comes from.

While looking at the second version of the Sauti article and trying to discover an overall frame, I wondered whether "the fight for democratization" or "self-governance" would fit the bill. However, these are not uncommon frames, especially in INGO-related media from sub-Saharan Africa. Moreover, these frames connect to the discourse of neoliberalism that has been present in the DRC since US president Kennedy's interference (Dunn 2003: 97) and perhaps, one could argue, since Leopold's. Thus the article, although showcasing agency and complexity, is not ideal in terms of its conscious use of framing to avoid Africanist discourse. Furthermore, it does not do the work it is supposed to do in introducing a new lexicon that can help with frame and discourse diversity in the public discussion around refugees.

In searching for a specific overall frame as *metapherein* to use in the Furaha article, I felt the FRIs could help; I had found them so valuable in the first place—surely I could mine their value further? The NVivo analysis revealed that the frame most valued by the interviewees was "our shared humanity." This was evident in their direct suggestions to me and in their responses to the two-minute reporting clips. For example, Graham said: "Anything to bring to light the human side of the experience is worthwhile, because journalists and consumers have to remember human beings. We're seen as numbers, statistics, rather than actual beings with dreams and hopes." Similarly, Maria said: "What I just want they should just talk more about those people having a life before." And Thomas, when discussing what he liked about one of the clips said: "I liked the way the journalist asked the question to find how was her life back home. This is a really human question."

In focusing on the "our shared humanity" frame for the Furaha piece, I am not claiming that it is the best or the only correct frame but rather that it wasn't a frame that kept alive Africanist discourse. In focusing on shared humanity rather than the humanitarian frame, it is also a frame that wasn't speaking to an already powerful discourse in the world of refugee camps. I hoped to structure a story where "their bare life" became "our shared humanity" so metaphor as *metapherein* would also be present.

Writing while practicing frame reflection was a completely different experience the second time around. After I had decided that I would focus on the frame of humanity, I found that I was able to sketch the bones of the Furaha story in a matter of 20 minutes on a scrap of paper. In this sense the frame returned to its function as a quick working routine for journalists. I was even having fun.

In terms of practicing frame reflection on a line-by-line basis, this also occurred. However, this time around I found it easier and therefore, importantly, quicker, to resolve tensions by referring back to the overall frame. Here is an extract from my first draft:

> Kahumbu knows that the rent free situation can't continue forever but she believes that she will have at least two more years to help her rebuild her life before the house is finished.
>
> *Quick community support, quick not feeding people, quick numbers down but those stayed (fine to do this quick obligation is to the frame not to the NGO interviewees!)*

At this point I had started to think ahead, planning where the article will go, structure-wise, through the use of links. The italics were a note to myself, the key being the repetition of the word "quick." The information that I was about to include next came from my interviews with English-speaking officials in the aid community. I had good, full quotations I could use. I wanted to use quotations. As shown by Hall in *Policing the Crisis* (1978), quotations from those with power are what make up the bulk of journalism articles. This is what I actually started to do but then stopped myself. I wrote out this note so I wouldn't slip again. I had to force myself to deal with these sections sharply. One of the key ways in which framing occurs is simply through the amount of space allocated to particular parts of the story. If I had used long, full quotations from these officials, I would have drowned out the short, bare quotations from the refugees. This would have changed the power dynamic and erased the shared humanity frame I was trying to build because it would have made the refugees weaker. These officials get their voices heard enough. I was trying here to have other voices heard as equals.

In the preceding example I believe I was able to undertake successful frame reflection while writing. I now felt comfortable that this process could be incorporated into ordinary journalism production. However, this process isn't foolproof. An early version of the article, which I sent off for publication, included a massive mistake:

> Every morning the straight-backed girl, at that awkward stage of growth on the cusp of womanhood, wakes early, and walks eight kilometers to the world heritage listed Virunga National Park, home to endangered primate gorillas and dangerous human guerrillas; the latter are violent militias who hide in the forest and have made international headlines for using rape as a weapon of war.

The underlined section is of course the mistake. This mistake came about for several unjustified reasons. First, I included the danger that Furaha puts herself in and hinted at the possibility of rape, because ever since it was dubbed the "rape capital of the world," that is *the* media story of the DRC. I was responding to the news values of "follow-up," "conflict," and "negativity" and felt this inclusion would help the story get published. Next, I was carried away with the seductive wordplay and alliteration of the repeated gorillas/guerrillas. Driven by the institutional action frame of readability, I completely overlooked the fact that I was invoking the animal trope of Africanist discourse. It was only after reading the words "Colonial and racialized scripting of the conflict in the DRC and its main players portrays them as continuing to reside in a bygone era where and when beasts rule the jungles" (Eriksson Baaz and Stern 2013: 26) that I realized what I had done. The article had already been sent for publication! However, it came back to me. I was told it could only be published if it was significantly shortened. I cut it down by 40 percent and I sent it again, this time with the offending sentence removed. The article was shortened still further and sent back to me for approval; the gorillas/guerrillas sentence had been reinstated. I was not surprised, it was seductive writing. However, I was determined to remove it. This time returning the article via email I wrote:

> In the edited version I had taken out the link between the gorillas and the militias. I did this because the association between the two is something which western foreign correspondents have been criticized for and it was wrong of me to include it in the first place. Unfortunately, carried away by the story flow I hadn't realized what I'd done until I'd read it again for editing.

Although my rationale is not detailed, it was obviously enough. The offending sentence was left out of the published version. Early in my research, I wondered if reporting from sub-Saharan Africa should be subject to a type of affirmative action to try to counterbalance the negative effects of traditional framings. In this particular case, the answer arrived at through practice was yes.

The experience of practicing frame reflection the second time round proved to be more successful, both subjectively in terms of my experience of writing the second article, and objectively in terms of the outcome of having the story published in a desirable publication. The idea of being published in *Crikey* was appealing because I know it is read widely by other journalists.[7] One of the reasons for attempting to publish an article, as opposed to simply creating it, was to expose other journalists to a different frame.

But publishing came at a price. I had to cut whole sections. The male voices in the story were sacrificed *even though* the absence of positive male representations was an important criticism that came out of the co-design process. And the published story focused more closely on Furaha, on one individual. This worried me.

On April 24, 2015, I wrote a summary of all my concerns on my blog. I worried that perhaps I made a mistake in focusing closely on one person's story—in making that the scaffolding of the article, I worried that I had rendered the frame, to use Shanto Iyengar's term, "episodic" (Iyengar 1994). Iyengar argues that news structures favor episodic frames—that is, stories framed around an event or a person. He contrasts this to stories that are more contextual, which he calls the "thematic" frame. Iyengar's research probed audience responses to the differently framed stories and found that the more commonly used episodic frame sees audiences ascribe the causes of poverty (at least in the US context) to the individual, whereas the less commonly used thematic frame tended to leave audiences looking to society or government for responsibility and/or solutions to the problem of poverty.

Furaha is an impressive individual, but she is one of many, and those individuals who are less "newsworthy" than Furaha, but still suffering from the same war, are just as important to relate to. The issue of refugees stuck in horrendous situations is not represented correctly if people come away believing that the story is only about heroic individuals who beat the odds:

> Getting the balance right between enough detail of the individual to create that relation to and yet not so much that the frame becomes episodic seems an incredibly fine line and I'm afraid sometimes while walking it I feel like I have two left feet. (Giotis 2015: online)

The edited Furaha piece is published and cannot be changed. Despite some concerns regarding the finished product, the experimentation proved to me, at least in this small pilot, that frame reflection could work and was a valuable process in and of itself. Moreover, there is potential willingness and consciousness in the profession, at least among some individuals, for affirmative action to remove Africanist tropes, and the FRIs are an extremely valuable practice. The published version, titled "What It Takes to Survive as an Orphan in a Congolese Refugee Camp" is available through the *Crikey* website.[8] This is not the published version, but these are the words that have been most fashioned by the FRIs . . .

Dateline: Goma, DRC, August 30, 2014

Furaha is 15 years old. Asked the worse thing about living in a refugee camp on the eastern border of the Democratic Republic of Congo, she, like everyone else, answers "the hunger." But then, in an intuitive and self-conscious gesture reminiscent of so many teenagers, she plucks at her old, ill-fitting T-shirt and says, "Also I miss having nice clothes."

Furaha is an orphan, war survivor, hardheaded entrepreneur, and average teenager.

Every morning the straight-backed girl, at that awkward stage of growth on the cusp of womanhood, wakes early, and walks eight kilometers to World Heritage–listed Virunga National Park. This battered tourist destination is periodically closed because of violent militias who evade capture in the forest and make headlines for using rape as a weapon of war. Yet Furaha is determined to collect sugarcane. Loaded to capacity, she walks back to her "home," "Mugunga refugee camp number 1" on the outskirts of the city of Goma in the eastern Democratic Republic of Congo (DRC).

She has no sign, stall, or cash register, but she has a regular spot where she lays out her wares—the perfectly round, streaked-green sticks that are sucked and chewed for sweetness and sustenance.

"I have no brothers," she says, in answer to the question why she walks 16 kilometers a day and sells sugarcane instead of going to school. Her parents and brothers were killed in the fighting that brought her to the camp, and now refugees in Mugunga are no longer being fed—so Furaha turned to a risky business model to survive.

Yes, she would rather go to school, but her dreams are modified to her reality.

She plans to move into the more profitable sale of cassava and potatoes; after that, she dreams of moving to the nearby town of Goma and becoming a successful entrepreneur there.

Furaha's story, the drying up of aid money and her risky struggle to survive as a small businesswoman, is no isolated phenomenon, and if the current trend for drawn-out, militia-based conflicts continues, her story will be increasingly repeated worldwide.

In the DRC the conflict on its colonial-drawn eastern border has been raging for over two decades.

The major war, involving nine nations, known as Africa's Great War, is over, but the peace agreement signed in 2003 means nothing to the militias who stake out land and resources, and ensure eruptions of violence that the world's largest peacekeeping mission, 20,000 UN soldiers strong, headquartered in Goma, can't eliminate.

This continuous conflict creates its own multifaceted reality.

"Goma is an IDP city," says Salomé Ntububa of Christian Aid in Goma. In the lingo of the United Nations and the International Non-Government Organisation (INGO) workers who proliferate in this town, IDP means internally displaced person, citizens, like Furaha, fleeing war, not over a border, but to a different region of their own country.

In 1992, when the end of the Cold War also ended President Mobuto's kleptocratic, US-supported reign, and the breakdown of Zaire (now DRC) began, Goma's population was estimated at 100,000. Today estimates top

1 million. Ntubaba points out that Mugunga camp was once considered part of the national park; now it is more an outlying suburb of the city.

It is important to understand the relationship of the city to the camp because an untold story of refugee and IDP crises around the world is the role of local populations.

Face to face with the suffering of their fellow humans, those living in towns near the refugee camps cannot ignore their plight—the human instinct of solidarity sets in.

Relying on the Kindness of Strangers

Christian Aid's Ntububa and others[9] estimate that only 25 percent of the refugees and IDPs in the immediate region are in camps; the rest are supported by the local community.

Kahumbu is another sugarcane seller but unlike Furaha she does not engage in the risk of collection. She's just a saleslady on the main road between camp and town. The widowed mother of five came to Goma at the same time as Furaha, during the M23 rebellion in 2012, and at first she too went to Mugunga camp.

"My children absolutely refused to stay in the camp," remembers Kahumbu.

The way Kahumbu tells the story, she hadn't much choice as her two tweens, in particular, remonstrated dramatically and Kahumbu herself was still shell-shocked from the violence. However, she doesn't regret listening to her children.

She found a stalled project; a half-built house and a builder-owner who has let her family stay there, rent free, for the past two years.

Kahumbu's rent-free situation can't last forever, but she will have two more years to get back on her feet before the house is finished.

How to better support people like that builder-owner who in turn has supported Kahumbu has always been the underacknowledged missing link, and it is now a question asked with urgency by the aid community in Goma—especially since the International Organisation for Migration (IOM) stopped feeding people.

Neighboring wars and wars further afield have budgetary priority. There is no longer a "general food distribution"; only those classed "most vulnerable" are given food. IOM head for the region Monique Van Hoof acknowledges the lack of funds was crucial to the food decision but also talks of finding "durable solutions," envisaged as community-based solutions.

Exactly what the international aid community can do remains to be seen but it has been adopted as official policy by the UNHCR for the DRC in 2015[10] and several INGOs are changing their practices to support "host communities."[11]

In the meantime with no food Mugunga camp is emptying. Some are heading back to decimated villages with limited support, others are carving their own path.

New Meanings to "Business Risk"

In the camp six men work furiously at a huge workbench cutting long pieces of wood into timber. They are sweating in the sun and refuse to stop for even five minutes as they have a large order—their business's first order—to fulfill.

So that I can conduct an interview my guide offers to take over on a two-man saw but soon finds his sedentary office life has left him unprepared for the work.

Laughing, 23-year-old Shukur explains that all the men were trained as carpenters in their village in the region of Masisi. Although able-bodied men were the first, many months earlier, to stop receiving food, the husband, and father of three young children, won't return and risk capture or death.

"I'm afraid to go back to Masisi because of the rebels," says Shukur. Right now, with the outdoor workshop, fine weather, and a big order, the decision to stay seems right. He may even fulfill his dream of having a fourth child.

"I would like to have four children, but no more, as I don't think I would be able to support more than four," says Shukur, as he resumes work, smiling, full pace.

"I feel good," puffs Shukur. "Of course I feel good because I'm going to receive something for my work."

Soon the planks will be ready to be delivered, on time, to the builder in Goma.

If Goma is a city of migrants, it is also city of enterprise. The city's landmark is a faux-golden statue of a tshukudu.[12] This is a wooden bike with an extra long and wide down tube perfect for transporting products through muddy, volcano-rubble-strewn streets and mountain villages. (Watch tshukudus racing here.)[13]

Tourists pose for pictures in front of the larger-than-life statue, the handsome driver carrying the whole world—a globe—on his tshukudu. And, indeed, minerals transported by tshukudus are in electronics all over our interconnected world.

In this enterprise-worshipping town one "durable solution" for INGOs is supporting small businesses; the logic being a strengthened economy means the city's population can do even more to help refugees and IDPs.

In a suburb of Goma a cooperative of cooking pot makers is part of this international experiment. Money has come from the west through Christian Aid, to small business lender PIAMF, then to this business of 15 pot-makers with an innovative pot design that uses less charcoal.

However, small business, especially in Goma, is risky business. The cooperative took a second loan after mass looting during the 2012 M23 rebellion. They recently signed the lease on a more secure place to store their pots—maybe just in the nick of time.

Even Riskier Times Ahead

2015 sees the DRC on a knife-edge with President Joseph Kabila and the UN at loggerheads over the details of military operations against remaining militias. There is the chance the UN mission will leave.

In a dubious honor the people of Goma would be happy to live without, the highly respected <u>International Crisis Group says the DRC is</u> one of the 10 wars to watch this year.[14] Obviously, this is not the situation everyday small businesspeople want to operate in.

Thirty-seven-year-old Chantal Samvura was a small businesswoman in her village. The confident, well-spoken woman produced and sold beans and charcoal—now, thanks to the destruction in 2012, she is one of the Mugunga IDPs. Samvura misses her village desperately; she misses the weddings on Saturdays and worries about the situation in the refugee camp where instead of dating and marriages, girls are abducted by boys. She also worries about her seven children's education in the camp, which she says is second rate compared to what they were receiving in the village. Partly this is because the children go to school hungry.

"Also, the teachers were much better in the village. We had very good teachers," says Samvura.

Samvura wants to go back to the village to start up her business again, that and a small orphanage which will more than likely be much needed, but she will not go back without some savings so she can replant her fields and without absolute certainty that it will be safe.

Her husband went back to see if it was safe and was killed.

Right now the widow with seven children will not risk cane-selling either. She prefers to keep her low-paying but fairly secure job as part of the sanitation team of the camp. Among other things she appreciates the position because it gives her access to soap so she can keep her family clean. In a place like Mugunga it is impossible to overstate the value of some dignity and an assured income.

Reliance on the idea of small businesses as the "durable solution" for these men and women ignores the fact that few will ever earn enough for more than basic survival and it ignores the precarious situation these businesses operate in, where some sort of buffer for bad times is not an insurance luxury but a necessity.

It also ignores the fact that not everyone is a born entrepreneur.

Sitting near Furaha's sugarcane a young mother named Sifa reveals she gave up on the sugarcane trade herself, finding it unprofitable. Later, Furaha is not on hand when a big sale comes along. All the cane is being bought at once, by my guide, to say thank you to the women for giving their time for interviews.

There is a wild, sudden, desperate scramble for the free, life-sustaining sugarcane. Only two toddlers, playing with pebbles in the dirt are immune, unaware of the importance of what has just taken place.

The rampageousness brings Furaha out of a nearby tent. She is furious that such a big deal has been transacted on her behalf. Immediately I shove another note into her hand; she gives me a split-second cheeky half smile, but to the women she continues screaming, yelling at them for their impudent and undervalued sale.

An hour and a half later Furaha is walking through the camp; in her hand she clutches a packet containing a cheap hair extension of the type seen everywhere in Goma. It's been a highly unusual morning and freed of cane selling for the rest of the day there is a bounce in the 15-year-old's step and a look of happiness on her face as she makes her way through the camp, fashion item in hand, and gets ready to spend some time on her hair.

NB: the author has respected the fact that some interviewees wished to give for publication their first name only.

PART II

WORLDS

3

The Space-Time behind the Storylines

In the swimming pool beside the neatly laid out tables at the Mamba Point restaurant in the Sierra Leone capital of Freetown that evening, white women were being taught water aerobics, and the conference hall near the entrance was hosting the seminar "The Traumatized Child." Mantovani's strings played around us, ice cubes tinkled in our glasses of white wine. . . . In "humanitarian territories," the restaurants, squash courts, and golf and tennis facilities are often back up and running before bombed-out schools and clinics.

—Polman 2010: 48

What the previous two chapters have shown is that the construction of words is exactly that, a construction. They come with their own histories, and it takes innovative methods and reflexive thought to find new words and new frames to tell new stories. This chapter seeks to change register. It is still concerned with the structures that might impact our word choices, but what this chapter focuses on, in greater detail, is the structuring role played by the social geography we inhabit. To understand the practice of foreign correspondence in borderlands it is crucial to first understand the dissonant social geographies that have been created thanks to the fact that "many countries now find themselves hosting large foreign contingents of donor representatives, UN specialists, aid workers, consultants, private contractors and foreign militaries. They have become the laboratories of the new liberal imperium that Michael Ignatieff has called *Empire Lite*" (Duffield 2007: 135).

The city of Goma is a singular place in the world, but it does not have a simple, singular identity. Goma can be thought of, and is, a Congolese city, the capital of North Kivu and the most significant city in the eastern region of the DRC. Goma has a population estimated at 1 million, and it has one major university. It is situated between the Rwandan border, only one kilometer to the east, Lake Kivu, its border to the South, and the still active volcano

Borderland. Chrisanthi Giotis, Oxford University Press. © Oxford University Press 2022.
DOI: 10.1093/oso/9780197565797.003.0004

Nyiragongo to the north. In 1994 in the aftermath of the Rwandan genocide, this small regional city of 100,000 people was transformed thanks to the influx of almost three-quarters of a million people fleeing Rwanda. With its thriving economy (much of it conflict based) and diverse population (diverse both in terms of Congolese cultural/ethnic groups, and in terms of hosting many internationals), Goma is significantly different to the surrounding countryside, and plays an important role in the region. A useful comparison has been drawn with London and the southeast of England (Rawlence 2012: 19). Goma is considered a safe city in terms of the surrounding conflict, as it is the base of the UN mission in the DRC MONUSCO (Mission de l'Organisation des Nations Unies pour la Stabilisation en République démocratique du Congo). However, in November 2012, in a show of strength and defiance, the M23 rebels captured Goma; they left of their own volition in early December.

Goma can also be thought of, and is, a humanitarian node created by war and by the happenstance of the city's existence on a border. Hundreds of thousands of people left Rwanda for neighboring Tanzania and Burundi in 1994, but Goma hosted by far the biggest contingent of Rwandan refugees. The Rwandans fled their country either because they were afraid of the killing or because they had been involved in the killing and now feared retribution. Goma is a reference point in the history of humanitarianism, breaking all records for fundraising and for the spatial concentration of aid organizations:[1]

> Before the Goma crisis, people were amazed if 40, sometimes as many as 80, international aid organisations arrived at the scene of a humanitarian disaster, but no fewer than 250 threw themselves into aid operations in the Great Lakes Region, along with eight UN departments, more than twenty donor governments and institutions, and an untold number of local aid organizations financed by foreign donors. . . . Never before had so many aid workers been gathered together in a single "humanitarian territory." (Polman 2010: 20–21)

The Goma refugee crisis of 1994 and the resulting cholera epidemic clearly showed up the interaction between humanitarians and the media. INGOs[2] competed with each other, using inflated atrocity claims and photogenic spokespeople to get on the nightly news (Dowden 2009: 7, 248). With some notable exceptions, both the humanitarians and the media ignored the politics of the situation, including the fact that former *génocidaires* were openly using the facilities of the camps (Polman 2010: 13–36; Dowden 2009: 248; Dowden, pers. comm., 2014). Rwandan president Paul Kagame justified his invasion of Zaire (which sparked the First Congo War and eventually the Second Congo War) as necessary to stop the *génocidaires* using the camps to regroup, build up their strength, and attack again. This peacekeeping mission is the "longest, largest and most expensive" in UN history, with a budget of almost US$1.5 billion for 2015–16 and more than 22,000 uniformed and civilian personnel deployed (Perera 2018: 7).

Finally, Goma can be thought of, and is, a place that journalists occupy. A place moved from material reality into symbolic reality via foreign correspondence. A movement that happens through choices based on generic conventions and material experiences . . .

Dateline: Goma, DRC, August 31, 2014

A little kid is looking in through the window of my hotel room. Welcome to Hotel Versailles, Goma, DRC.

How should I capture Hotel Versailles on the page for you? The journalist in me wants to tell you the dramatic story of my first night. Hearing people walking past my door, glimpsing, through the small crack in the drawn curtains of my closed window, a backpack, a backpacker's backpack, and knowing that another young westerner, probably, like me, following the *Bradt Guide*, was staying at the hotel—would be staying next door in fact. By the size and style of the backpack I guessed a guy.

Then, later, but not much later, the knock on the door, not my door, the backpacker's door. The word: "Police." More conversation I couldn't hear properly, maybe it wasn't English, maybe it was French.

Before I even knew I was doing it, I found myself pulling table and chair against the door to create a flimsy barricade, wishing, oh how I wished, that I had taken the advice of the international security company, paid for by my university, and packed a rubber doorstop. Heart pounding, awake, alert. Vowing to stay awake. Eventually falling asleep from nervous exhaustion.

But that story—that drama—would be, not a lie, but certainly a misrepresentation of the whole.

Much more representative of my time at Hotel Versailles is this little kid, I don't remember his name, looking in through the window, the window that

I now almost always keep open, standing there, mute but wondering, while I Skype Steve in Australia. Me doing introductions: "Hello, meet my husband who lives in a computer"; the joke would have worked better if I could speak French, Swahili, or Lingala. Silence from the kid, and then an awkward wave in response to Steve's enthusiastic one. I got the feeling he didn't really want to be forced to be part of the Skype call; he was just curious about the strange noises coming from my room and so came along to have a look.

Hotel Versailles was like that—the space encouraged a lot of communal interaction. It was a hotel where people stayed for long periods of time and it was part of its neighborhood.

You enter via a wide gate, which is only shut at night, into a courtyard and parking lot. To your left is the small administration building, and directly in front of you, at the back of the compound, is the building itself. Visualize, if you will, two stories, wide and low. The basic design is similar to the American motels I've repeatedly seen in movies except, unlike those motels, this hotel looks elegant in a simple, unglamorous, slightly run-down, old-world kind of way. The hotel is painted white and black and each room has its own small courtyard demarcated by low Greek columns, like those seen in suburbia on balconies. For me, with my Greek Australian background, this detail means it is both elegant *and* homely. Beyond your little demarcated courtyard is the common walkway. On the top floor, the walkway ends amid more Greek columns and the veranda from which the richer guests survey us. But on the ground floor, my floor, there is a small, grassed area used to dry the hotel guests' laundry. In the middle of the lawn is a small palm tree and then a wall made of beautiful volcanic rock. This is the privacy wall separating the private hotel spaces from the public hotel spaces on the other side, the bar decorated with bamboo and palm fronds, and the lovely garden with tables and beach umbrellas bearing beer logos. The bar will come back into the story later—after all, I am a journalist. We are known for our frequenting of hotel bars.

For now, let me tell you about the large water tank and how it becomes a lifeline for other shopkeepers nearby on the too-frequent occasions when water in this neighborhood suddenly refuses to come out of the pipes. Finally, there is a large parking lot filled every night with police and army four-wheel drives and jeeps. The cars are collected each morning. I don't think that all the owners are staying at the hotel; I think that the cars are there because in the parking lot they are protected by the compound walls, gate, and security guards at night. However, several police officers are staying at the hotel, including someone who I am able to identify as a high-ranking officer. Perhaps he is conducting some sort of training. It is his kid, the only kid in the hotel, who comes to my window, and I am in awe of his wife who is always beautifully dressed, always perfect in manners, always friendly and welcoming to this strange western girl who doesn't speak French, Swahili or Lingala. If this were an Agatha Christie novel, she would be the character described as a "true lady."

Needless to say, as I get to know this family a bit better, I start to feel sheepish about the barricading episode. After all, it might have been this man at the door that night. I begin to think maybe he had just been introducing himself, or maybe just checking up. Mercenaries are certainly not unknown in the DRC; potentially he had just been doing his job and doing it well. I was never able to find out about this incident—even if I'd had the guts and opportunity to ask the police officer, I wouldn't have had the language skills; and I never had the chance to get the western guy's tale because it turns out he didn't stay the night. That very night he moved to a more expensive hotel catering to westerners. I never discovered who he was. I only heard about his move some days later from a tourist tout—we westerners get watched.

After that first night, and for the 10 days that I stayed at Hotel Versailles, I was the only westerner.

Thank you Hotel Versailles for the humbling and enlightening experience.

Space as an Analytic Tool in Journalism

Space is fundamental in any sort of communal life; space is fundamental in any exercise of power.

—Michel Foucault in Mahon 1992: 30

In 1984 the death knell sounded for the foreign press corps in Beirut. While in the past they had been the focus of irregular threats or even shootings, by 1984 they had become primary targets for violence. *Symbolically, the Commodore Hotel's bar*, the great gathering place for the foreign press, was trashed by a Shiite militant.

—Lederman 1992: 121; my italics

In the literature on and by foreign correspondents, the space of their hotels often makes an appearance. In adventurous "autobiographical tales of derring-do" (Murrell 2009: 7) the hotel is a site of refuge. Hotels offer a break in pace from the drama but also, as a literary device, they function, as in real life, as a space allowing reflection, contemplation, and confession, where the heroic facade of the foreign correspondent can be laid aside and time is taken to consider the complexities, foibles, and contradictions of the experience, as when Blaine Harden writes: "Back in hotels in Addis Ababa, the Ethiopian capital, legions of First World famine-watchers—reporters, aid workers, politicians, actors, rock musicians—dined together over discussion of the advisability of repeated shampoos to get germs out of our hair" (Harden 1991: 12). Or when Michael Buerk, again in Ethiopia, returning from a village beset by famine, says:

> After that terrible day in Korem we got back to the Hilton in Addis. I was grateful for its luxuries but resented them. I could not see how it could exist in the same world, let alone the same country. . . . There was a huge swimming pool in the shape of a cross, fed by a warm spring. In the immaculate gardens there were tennis courts and a pitch and putt area for keeping up your golf. Inside, there were saunas and massage parlours and squash courts. There was a pastry shop as well as the bars and restaurants, and they were all full of people eating and drinking; people who took public money to stop Ethiopians starving. I was overwrought and full of disgust. I had my own sense of guilt to cope with, too. (2004: 288)

In descriptions *of* foreign correspondence, hotels play a different, harsher role—they are metaphors of critique. The term "hotel journalism" entered mainstream discussion during the Iraq war, with reporters acknowledging that, although they signed their reports with the dateline "Baghdad," in fact they had not been out of their hotel all day, which was "a virtual high-rise bunker" (Goldstein 2007: 84; see also Fisk 2005). This "hotel journalism" was, and is, considered necessary due to the extreme risks posed to western journalists on the streets of Iraq. The pros and cons of such journalism is still a topic of critique and debate. Harris and Williams discuss war correspondents as particularly prone to being "hotel warriors" (2018: 65), which can morph into pack journalism, meaning the loss of independent reporting, the narrowing of sources consulted on a story, and a herd mentality that examines only one dimension of a topic: War reporters are seen as particularly prone to the phenomenon. The hostility of the environment in which they operate accentuates the normal inclination of journalists to socialise and cooperate with one another in foreign parts" (Harris and Williams 2018: 66). My own interest in talking about hotels falls into neither of these camps but borrows from both. Hotels are places of reflection and escape, and hotels are also spaces of work where a lot of journalism gets done. What I hope to add to the discussion around hotels in the journalism literature is an analysis of how hotels, and other "safe" expat places like bunkered residences, bars, and cafes, function as part of the apparatus that creates an internationalized zone of being in Goma. INGO-/UN-dominated spaces, not only in Goma, but in conflict zones around the world, have been ironically dubbed "Aidland" (Mosse 2011a) or "Peaceland" (Autesserre 2014) and critiqued as a " 'lala land,' a developing country which has no history, geography or culture" (Althorpe in Wallace, Bornstein, and Chapman 2007: 173). This is the space in which foreign correspondents operate; it structures the way they conduct themselves and the way they are seen by locals and it exerts *affects* that impact the journalism they produce. My argument is that spaces are important. If they weren't, there would be no discipline of architecture and we wouldn't care about shows like *Grand Designs*. But it is one thing to acknowledge space; journalists are expert at this—we feel space and incorporate it into our narratives. It is another thing to analyze space in such a way as to contribute to the critical reflexivity of the profession.

Chris Nash in *What Is Journalism?* argues that journalism is transitioning to a "methodologically self-aware, critically reflexive practice in the production of knowledge" (2016: 9). One of the ways this is happening is through analysis of journalism as deeply embedded in spatiotemporality, with both space and time understood as being socially constructed. Nash argues that "spatiotemporality is widely recognised as a fundamental issue for journalism practice" (2016: 28). He references key theorists looking at journalism production, including G. Stuart Adam and James Carey, as detailing the importance of journalism knowledge production as a process embedded in specific spaces and places. Further, Nash draws on sociologists Gaye Tuchman and Richard Ericson, key theorists looking at the relationships between journalists and their sources, to argue that "relationships between journalists and their sources take place in material contexts, with spatial, temporal and organizational characteristics that constitute highly specific terrains for each encounter in the relationship" (2016: 186). Throughout chapters 3, 4, and 5 I will show how Goma is a highly specific terrain that creates its own relationships between journalists, their sources, and more broadly the culture and society they are purporting to represent.

In the organizational characteristic of each encounter noted by Nash we see a concern similar to that raised by Schön and Rein: "Even a chat between close friends occurs in the institutional setting of someone's house or a walk around the park" (1994: 31). We have already observed the impact of these institutional settings in terms of "frames of reference," with the interview space of the refugee camp affecting the topics under discussion.[3] By moving the focus of analysis from frames to space, it is possible to develop a deeper understanding of these issues of institutional context. The overall aim is to intertwine these two analyses and view the multiple forces impacting upon the mode of foreign correspondence taking place in spaces like Goma.

A theoretical framework advocated by Nash, and which I will also use, is David Harvey's spatiotemporal matrix (Figure 3.1) in which Harvey

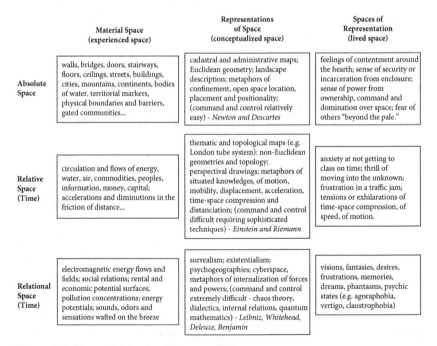

	Material Space (experienced space)	Representations of Space (conceptualized space)	Spaces of Representation (lived space)
Absolute Space	walls, bridges, doors, stairways, floors, ceilings, streets, buildings, cities, mountains, continents, bodies of water, territorial markers, physical boundaries and barriers, gated communities...	cadastral and administrative maps; Euclidean geometry; landscape description; metaphors of confinement, open space location, placement and positionality; (command and control relatively easy) - *Newton and Descartes*	feelings of contentment around the hearth; sense of security or incarceration from enclosure; sense of power from ownership, command and domination over space; fear of others "beyond the pale."
Relative Space (Time)	circulation and flows of energy, water, air, commodities, peoples, information, money, capital; accelerations and diminutions in the friction of distance...	thematic and topological maps (e.g. London tube system): non-Euclidean geometries and topology; perspectival drawings; metaphors of situated knowledges, of motion, mobility, displacement, acceleration, time-space compression and distanciation; (command and control difficult requiring sophisticated techniques) - *Einstein and Riemann*	anxiety at not getting to class on time; thrill of moving into the unknown; frustration in a traffic jam; tensions or exhilarations of time-space compression, of speed, of motion.
Relational Space (Time)	electromagnetic energy flows and fields; social relations; rental and economic potential surfaces; pollution concentrations; energy potentials; sounds, odors and sensations wafted on the breeze	surrealism; existentialism; psychogeographies; cyberspace, metaphors of internalization of forces and powers; (command and control extremely difficult - chaos theory, dialectics, internal relations, quantum mathematics) - *Leibniz, Whitehead, Deleuze, Benjamin*	visions, fantasies, desires, frustrations, memories, dreams, phantasms, psychic states (e.g. agoraphobia, vertigo, claustrophobia)

Figure 3.1 Harvey's Matrix of Space and Time
Source: Harvey 2006: 135.

combines his own theories with those of philosopher and sociologist Henri Lefebvre. This matrix uses real-life examples to help us see the different conceptions of spatial temporality and spatial relativity available to us, and how these manifest.

Nash suggests using the matrix as a metatheory in journalism research that can be combined with the disciplinary concerns of journalism and the disciplinary concerns of the field under investigation.

> It is a meta-theoretical checklist that identifies what sorts of dimensions need to be considered when dealing with any given phenomena in the world. Underneath meta-theory will sit the theory specific to the specified phenomena, e.g. climate theory and practice for climate change, military theory and practice for war, medical theory and practice for health . . . it gives a structure to reflexivity, and also points to the interdisciplinarity that is a core aspect of journalism. (Nash 2016: 118)

In use of the matrix Harvey emphasizes that quadrants should not be looked at in isolation:

> The only strategy that really works is to keep the tension moving dialectically across all positions in the matrix. This is what allows us to better understand how relational meanings (such as value) are internalized in material things, events and practices (such as concrete labor processes) constructed in absolute space and time. (2006: 147)

Here, then, is a further overlap with frame theory, for in looking through material space we can understand frames of thought that have helped create those spaces and places. This is clear in the middle column of the matrix, where Harvey recommends we consider metaphors of situated knowledges as a key way of understanding how conceptualized space works.

I used the matrix to help structure my reflection on my practice of foreign correspondence in Goma. It helped me think through the repercussions attendant in the meeting of particular professions, with their own disciplines of thought, in a particular space, with its own competing and complimentary logics. I will present that matrix soon. But first, to better understand the context, to set the scene if you will, a few more sketches. These sketches are of me in Goma, of my Goma . . .

Dateline: Goma, DRC, August 24, 2014

Let's start again at the Hotel Versailles. I promised you I would come back to the restaurant/bar.

It's late afternoon on a weekend and today the beer garden is full to overflowing, music is playing, and people are having a good time. People are drinking. I am sitting at a table in the garden by myself. I have ordered fish head stew. The handsome waiter, who is usually extremely grouchy, is pleased that I am trying a traditional dish. It is taking a long time to prepare because it is being made fresh from scratch, but that's OK, I have plenty of work to do and I am tapping away on my laptop. The night is young and I like Congolese music. Actually, it is one of the only instances when not knowing the language may be an advantage. I enjoy the rumba rhythm—a music that was once, in the 1960s, the soundtrack to freedom—in those heady days before it was decided by the CIA that the DRC's first president, Patrice Lumumba, must be removed. However, the modern lyrics, I am told, are peppered with paid-for inclusions of politicians' names—I am sure this would have annoyed me.

Suddenly, a young man is standing at my table. He solemnly declares that he loves me. "I love you," he says over and over again in English. He is drunk. I am pleased he is standing and not seating himself, but equally, he is not going away. Eventually two of his friends come and collect him with "What can we do?" grins of resignation and amusement directed at me, but my suitor is still enamored, and periodically returns to declare his love. If he knew the word "undying," I am sure it would be his un-dying love.

Now I wish I hadn't ordered fish head stew.

Eventually, with broken English and hand gestures, I tell my waiter I will eat in my room. I have never done this before, but I have seen other guests (admittedly those on the top floor) have food delivered to them. My waiter is not happy anymore. I am not sure if he is not happy at the situation with the drunk guy or at having to deliver to my room.

When he arrives I realize we have a problem. The stew is aromatic. If I eat it in my room the whole place will stink; he doesn't know where to put it. But I bring my chair out and ask him to put the stew on the ledge of my little Greek column fence. Now he is happy again and when I take a huge sniff of the stew and say, "Yum, yum, yum, yum" with gusto, a wide smile comes to his face he responds with an enthusiastic French-Congolese "Weh!" Well you may think it strange, but I love this memory, and so I should. It was a moment of connection.

Dateline: Goma, DRC, August 29, 2014

This was a very different night.

That day I had one of those experiences where you can't help but make your life as a foreign correspondent sound dangerous and heroic. Any story that starts out with the words "I was arrested" and is placed in an African country is doomed to stereotype. And yet to be detained by police in the DRC is an everyday event. Our car was impounded until we paid a so-called fee for not having some particular paper. The whole drama was just about trying to get the "fee" as low as possible. It was even enjoyable, especially when watching some of the other impoundees, locals, railing and refusing to pay the "fee." The most difficult things for me were (a) being detained so long that it became impossible to get to Mugunga camp that day (b) emotionally reliving a much more serious arrest event from my past—but more about that in chapter 4.

The crux of tonight's story is that the drama of the day left me craving company, English speakers with whom I could digest the day's happenings. I reached out to a local INGO worker, a wonderfully helpful and generous young woman. She couldn't have dinner with me because she was hosting a monthly bilingual trivia night. But I should come along to that, in any case there would be people there I should meet.

This conversation happens in fits and spurts over text messages. At some point I have given up on western company for the night and ordered dinner from the hotel. It is not the handsome but grouchy waiter but another, younger, less confident waiter with no English whatsoever. I cancel my order, trying to explain I am going out with friends instead, hurry to my room to change, and run out of the compound. I have the name of the place but nothing more. I am assured that everyone will know it. As usual there are motos (motorcycle taxis) waiting at the gate. One of the drivers seems to know the place. He gives me a price. I have no idea if it is too much and tonight I don't care. I hop on and we're off.

As mentioned I have no idea where this place is. I don't even know if it's a hotel or a restaurant, but I am a little surprised to see that we are starting to head out of the center of town. This trip seems to be taking a while. I look left and right as I try to get some bearings while hanging onto the back of the moto. Now we are out of electric light range, it's dark, that thick darkness near the equator. Oh man. And now we're heading down a corrugated dirt road. Shit. The bumping feeling from the corrugated road travels to the pit of my stomach, lifts me, and slaps me back down on vinyl. Suddenly, funnily enough, I'm calm. It is the corrugated dirt road, that very specific feeling of being on a small motorbike on corrugated dirt—I've been momentarily transported through time and space to a previous life as Indigenous affairs reporter in Dubbo, another overly textured place where stereotypes of black and white Australians reign. Operating outside norms, I would sometimes spend Sunday afternoons on my little scooter just riding around the rundown housing estate on the edge of town. I know that visiting the estate is outside the norm because one of the greatest compliments I have ever received came in the form of a letter written by a newspaper reader, an elderly lifelong Dubbo resident, who told me that one of my articles inspired her to visit the estate for the first time in her life. Soon after she became a volunteer at the youth center there.

One Sunday, on one of these jaunts, somehow I ended up on farmland abutting the estate. I could never forget that first feeling of a 50cc scooter on a road built for tractors.

And now I'm loving being here. The night air whipping across my face, I am outside the compound walls and away from concerned conversations about safety at night. No one knows where I am—least of all me—and I am intoxicated by the freedom. Riding through the night on the back of this little bike, this is the most fun I've had in ages.

I'll shorten the rest of the story. I am taken to an abandoned hotel on the lake. It is the wrong place. There had been recent name changes. Luckily some strangers we ask know the place I want. It turns out to be just around the corner from Hotel Versailles! Walking into the large pizza restaurant I am overwhelmed by how packed with people, crowded around circular tables, this huge space is. I see the INGO worker who invited me already emceeing. Thankfully, I recognize someone else, a consultant who, like me, has no regular office space and often works from the expat coffee shop. I plonk myself at his table where he sits with two other people, a British researcher and her Congolese fixer, and I enjoy listening to the trivia questions on international news, the questions delivered in both French and English. At a break in proceedings my friend comes up to me; she is pleased I am seated where I am: "These were the people I wanted you to meet!" In retrospect, and as will become clear in the arguments to follow, this is something more than coincidence—the INGO/expat world is structured to work like this.

Back in the hotel, late that night, I find the waiter and receptionist have waited up for me. My request to cancel the food order had not been understood, and my dinner has been kept for me. I pay for the meal and tell the waiter he can eat it. He is elated. I think he was expecting a disagreement and instead he got a free meal. His delight at the cold food brings home another reality. It has certainly been a day of contrasts.

The Effect of Affect

We have spoken too much about consciousness, too much in terms of representation. The social world doesn't work in terms of consciousness; it works in terms of practices, mechanisms and so forth.
—Bourdieu in Bourdieu and Eagleton 1992: 113

My argument is that the affective life of individuals and collectives is an "object-target of" and "condition for" contemporary forms of biopower.
—Anderson 2012: 28–29

There is a point to giving you snippets from Goma, snippets of my body moving through the internationalized corridors of the cityscape. My concern with the bodily practice of western journalists began with my following the critical literature on the working relationship between foreign correspondents and aid agencies. Of particular interest to me was the concern, especially prevalent in discussions of sub-Saharan African foreign correspondence, that western journalists are effectively "embedded" with aid agencies (Dowden 2009: 4–9; Franks 2010; Franks 2013: 133–60; French 2005: 59; Rothmyer 2011), and that this may entail "similar trade-offs to going on location with the military" (Franks 2010: 79).

In journalism literature, the critique of this embedded practice is that foreign correspondents "rely on NGO 'ex-pats' working overseas to provide perspective" (Moeller 2006: 188). The results, particularly for Africa, are the perpetuation of negative stereotypes that correspond with the funding priorities of INGOs (Dowden 2009: 7; Franks 2010; Hannerz 2004: 134; Rothmyer 2011) and because of time, space and funding constraints a failure to cover other stories (Rothmyer 2011). Furthermore, as discussed in chapter 1, through their role in disseminating transactional frame stories, western foreign correspondents become "a cog in the world's humanitarian machine" (Balzar in Hannerz 2004: 46).

Important work has been done by journalism scholars looking at the wider dynamics of increasingly close professional relationships and norms between NGOs and media organizations, which are visible in the direct influence of PR strategies on the production of news items and the influence of journalism norms on INGO publicity strategies (Powers 2018; Wright 2018). In development and international relations literature, scholars speak of a media-aid complex (Perera 2017; Laudati and Mertens 2019) created through the entanglement of the two fields. INGOs are aware that media exposure influences both funding and policy outcomes, but at the same time media are also reliant on INGOs for content. This means that priorities shift in both fields (Autesserre 2014). Once again the DRC acts as an exemplar, with scholars highlighting that the symbiotic relationship between journalists and INGOs leads to the simplification of storylines with attendant, unintended harmful consequences and the narrowing of public discourse. The attention these select issues gain (in the DRC's case particularly rape victims and conflict minerals) means that funding flows to those particular problems and it then becomes hard for INGOs to change their focus. Both professions feed off each other to become trapped in narrow storylines (Autesserre 2012; Eriksson Baaz and Stern 2013; Ramalingam 2013: 32–35).

I agree with these critical scholars, and in addition, after my own reporting trip to the DRC, I developed a set of related concerns. I came from my fieldwork with an overwhelming sense of despair at the expansive and entrenched UN-INGO structures in Goma and their creation of a world of their own, complete with cocktails by the lake every Wednesday. This "Aidland" (Mosse 2011b) and "Peaceland" (Autesserre 2014), in my view, interacted with the underlying local culture in highly structured and autocratic ways. I had previously come into contact with this world, notably in Juba, South Sudan, where I had found myself listening in, over a frankly excellent dinner in a Greek-Sudanese restaurant, to aid workers' conversations on the best first-class flight lounges in the world, and in Ethiopia, again over an excellent dinner, to conversations centered on the best postings for night life (Bujumbura, Burundi, in 2011, for those wondering). Also in Ethiopia, one could see the flash UN offices and the way the Hilton hotel was occupied

as an office space. However, aware of the criticisms of embeddedness, I consciously sought to limit my exposure to this world.

This time, as part of my autoethnographic research, and as part of my
compliance with the university's risk management strategies, I consciously
threw myself into the INGO community, and I found the experience and, in
particular, feelings of disconnection from "real local life" discombobulating
and overwhelming. This feeling was heightened when, halfway through my
trip, I left Hotel Versailles and moved to an expat-oriented hotel. This move
was motivated by the wish to experience the impact of different spaces and
I certainly felt a difference. That difference can be theorized under the concept of affect:

> Affect is an impingement or extrusion of a momentary or sometimes more
> sustained state of relation *as well as* the passage (and the duration of pas-
> sage) of forces or intensities. That is, affect is found in those . . . resonances
> that circulate about, between, and sometimes stick to bodies and worlds,
> *and* in the very passages or variations between these intensities and
> resonances themselves. (Gregg and Seigworth 2010: 1)

So, in the intensity that passed from the bumpy ground on that corrugated
dirt road in Goma, through the moto taxi and then to my stomach, there was
affect at work, and it changed the state of relations. I had been afraid of the
dark, of the long journey, and of my driver, and then I no longer was. The fact
that this affect was a resonance of a sensation from years ago is a perfect example of the statement that affect sticks to bodies or worlds.

> Sigmund Freud once claimed, in his very earliest project, that affect does not
> so much reflect or think; affect acts. . . . However, Freud also believed that
> these passages of affect persist in immediate adjacency to the movements
> of thought: close enough that sensate tendrils constantly extend between
> unconscious (or, better, non-conscious) affect and conscious thought. In
> practice, then, affect and cognition are never fully separable. (Gregg and
> Seigworth 2010: 2–3)

If affect and thought are never fully separable, and if the power of affect is
on our actions, then to understand the field of foreign correspondence we
must also *feel* the field and analyze those feelings. My need to talk to other
expats after a difficult day, my enjoyment at the trivia night, including the

opportunity to network—these are the types of affects that are targeted by the securitization discourse. That affect may act upon a correspondent without us being reflexively aware of the fact is inevitable but problematic. A profession dependent on understanding and communicating the space around them must consider how their own passage through that space is impacting the knowledge they are generating about that place.

Ben Anderson (2012) discusses securitization as a modern form of biopolitics and this is what I seemed to have experienced in the borderland of Goma. Certain affects are enabled, as the attempt to control our bodies is an attempt to reduce and act upon the affects in our worlds. However, Anderson also sees a way out of this situation. He draws on various theorists (notably Antonio Negri and his collaborators) who have conceived of biopower as containing within it a power from below that has the possibility to fight biopolitics as power from above. He notes that affects are not reducible to the biopolitics of securitization. Affective relations and capacities have the potential to be escapes "*events*, ruptures and beginnings that herald the birth of new ways of living" (Anderson 2012: 35). My initial fear on the moto was the direct result of biopower as biopolitical securitization: the many discussions, backed by INGO and university policies, that view traveling at night by moto as dangerous and prohibited. My subsequent enjoyment of the moto ride was an affective escape from the biopolitics of securitization through the potential in my own body.

On the drive back to Hotel Versailles after the trivia night, some well-meaning new friends urged me to switch to a more expat-oriented hotel. Their advice was a natural part of their privileged existence in the Aid/Peaceland of Goma. As Wacquant (1995: 65) points out, "Specific social worlds invest, shape and deploy human bodies." Wacquant conducted his field work in boxing gyms. He found the social aspects of training together, knowing that others were doing the same as you, helped justify the problematic and difficult practice of stepping into the ring. This is why the feelings of warmth, camaraderie, and safety that I experienced at the trivia event, and then in the car, contributed to the appeal of their argument—moving hotels, I knew I would be cocooned in this social and cultural support. However, I was also exposed to counterarguments in the form of affective escapes—moments like my experience on the back of the bike, or the cross-cultural communication over fish soup. Analyzed in relation to Aid/Peacelands, these were affects arising from attempts to escape the "apparatuses of security" that aimed to fix me in a "specific spatial-temporal topology" (Anderson 2012: 34) of the expat lifestyle. Moments outside the Aid/Peaceland world allowed me to see it more clearly.

On my return to Australia, in an effort to make sense of the experience, I delved into the development literature and found a vein of fairly recent yet influential critiques concerned with the bunkerization of development professionals (Duffield 2012; Egeland, Harmer, and Stoddard 2011; IRIN 2011) and the effects of the Aid/Peaceland spaces and structures on the types of knowledge being produced and privileged (Autesserre 2014; Duffield 2014; Jackson 2005; Mosse 2011b; Apthorpe 2011; Eyben 2011; Harper 2011). This literature will be discussed in more detail presently. I also began a more detailed consideration of the value and values of foreign correspondence, considering which values (or aspects of values) it should discard and which it should embrace.

I brought all this together using Harvey's matrix, which gave a structure to my reflexivity. The objective act of plotting my lived experiences me in Goma, my Goma, into the matrix combined with the conceptual act of considering how those practices could help or hinder the development of decolonized foreign correspondence revealed much about the power structures I experienced and contributed to. Among other important points, the matrix labels and analyzes affects experienced; through that process, affect becomes conscious emotion, and then when narrated into journalism it "becomes collective and potentially political" (Wahl-Jorgensen 2016: 129). My first matrix, which I drew in chalk over several days on the shared driveway of my home in late 2014 while walking between the squares (thank you neighbors for your patience!) is reproduced in Table 3.1. The analysis created while thinking (and physically walking) through the squares, combining the experiences of the past, and the vision for the future, helped create the "critical and dialectic method" which Harvey speaks of as "vital to understand not only where we have been and how we have been re-made but also to understand where we might go and what we might collectively aspire to become" (Harvey 2006: 89).

A Matrix of Appreciation

> The real voyage of discovery lies not in seeking new landscapes, but in having new eyes.
>
> —Marcel Proust

Table 3.1 My Goma Matrix

	Material space (experienced space)	Representations of space (conceived space)	Spaces of representation (lived space)
Absolute space	Restaurant and bar deep inside hotel (Caritas guesthouse) restaurant and Bar fronting street (Hotel Versailles) Space of car (inside, separated) space of moto (exposed, shared) Lake with corpses Unlit and unpaved streets NGO cement	DRC map—Goma represented, nearby towns not shown Bunkerization Lakeside living (Goma as offering lifestyle benefits) Agricultural breadbasket ("Garden of Eden" Massissi region destroyed by conflict) "NGO cement"	Elation at sound of cistern filling = water is back again (Hotel Versailles) "Strange Muzungu" being watched / talked about (Hotel Versailles) Laptop alcove for working in companionable silence at night (Caritas Guesthouse)
Relative space (time)	Travel by car (speed-distance compression, crowding out other users on narrow roads) Length of stay Budget for stay (money) Circulation of goods, people, water, electricity, and information at hotels: Versailles vs. Caritas vs. Ihusi Visit to local post office Visit to Mugunga	"Muzungu" "Expat" Expert flying visit DRC Level 5 risk rating (used both for government travel advice and to calculate "risk" pay loadings) Online cartoon map of Goma focusing on expat themes like the boulangerie	Thrill (fear and exhilaration) of unknown Anxiety at no electricity/internet to file stories (Hotel Versailles) Happiness in interactions with staff and local shopkeepers (Hotel Versailles) Frustration in interactions with staff (Hotel Caritas) Frustration at "NGO speak"
Relational space (time)	Business relations Social relations Conversing Energy potentials Sounds Vistas My translator and his fiancée waiting for me at the door of the boulangerie with the security guard	Psychogeography of Goma Cyberspace representations of self in Goma to friends back home, and their responses Heart of darkness Capital city of East Expat curfew Discourses of development and securitization	Claustrophobia Visions of Goma as place to be afraid of going to Visions of being in Goma as "courageous" part of work Desire for socializing Desire for "fun"

The points in the matrix I created are not meant to represent a "truth" about the space of Goma. Instead, what is revealed, is the way in which, at any point in time, we "are inexorably situated in all three frameworks simultaneously, though not necessarily equally so" (Harvey 2006: 128). We can choose to give primacy to one spatiotemporal conceptualization (or way of understanding the world) over another—and often this is done without realizing that this is what we are doing. However, the fact remains that "what we do as well as what we understand is integrally dependent upon the spatio-temporal frame within which we situate ourselves" (Harvey 2006: 128)—in other words our journalistic choices are impacted by the spaces we occupy in complex ways that may escape our notice. To illustrate its use, I will home in on three examples.

The Hotel Buffet

The middle left cell draws our attention to the circulation of goods, people, water, electricity, and information and thus the different affordances of different material spaces. You will meet different people in different hotels and you will have different access to different types of information, all very important points to consider in foreign correspondence and a clear case for using space consciously as a tool.

However, when considering the relative circulation of goods a complex tension emerges between the camps and the hotels, with this tension heightened the more expensive the hotel is, and the more quickly one flips between the two material realities. What does it do to you to be among hunger in the morning (the desperate scramble for sugarcane) and waste in the evening (the all-you-can-eat buffet)? This tension is important to understand, as it seems to be implicated in the push to the transactional frame of journalism discussed in chapter 1. Take, for example, the following statement by legendary Australian war cameraman David Brill:

> The first thing that comes to mind when I think of the moral issues involved in the job is filming people who are suffering. Sudan is a very poor country but we journalists and cameramen might be in the Hilton having a beautiful lunch off the buffet. Then we get in the Toyota Landcruiser and travel fifteen minutes up the road to a refugee camp and children are dying in front of us from starvation. That is when I feel guilty. "Where is the balance? How do

you justify filming this?" Then I have to think, "I am here and I have to eat and I like to have a clean hotel room. I can justify my work and hopefully the film will help stop the suffering—*organisations like the UN will see it and do something*." (Brill in Leith 2004: 58; my italics)

This correspondent is obviously feeling a tension between the relatively lavish lifestyle provided for him to report on suffering and the lifestyle of those being filmed. The analysis of these tensions becomes really interesting once we move into the relational space and discern different values at play and not in play. In the resolution of the tension via the idea that "organizations like the UN will see it and do something," the value espoused, no doubt influenced by the other people sharing the buffet, is humanitarian intervention, which as discussed in chapter 1, is a value aligned with the white man's burden trope and not aligned with the idea of a demotic global public sphere. This analysis of material contexts allows us to see that staying in a place like Hotel Versailles shifts that dynamic. This is because coming back from the camp and sharing the hotel with other, more ordinary, people from the DRC reinforces the fact that the world is not simply divided into victims and saviors—a different justification for your work must be found and the trans-actional frame may be replaced with something more complex. Alternatively, the correspondent may come back to the Hilton hotel but shift their frame of reference from absolute material space (I need food, and refugees need food too) to the relational frame (what values are embedded in humanitarian intervention if that intervention supports a Hilton hotel 15 minutes from a refugee camp?)—this again could disrupt the transactional frame—in fact, it starts a whole new line of investigation.

Claustrophobia versus Corpses

Claustrophobia is one of the psychic states that Harvey gives as an example of the bottom right cell in the matrix (lived, relational space). This cell highlights the relational aspect of space-time, which means that the same experience may be lived differently dependent on "where peoples' heads are at" (Harvey 2006: 128)—an example Harvey provides is the different ways in which different people will experience the same lecture given by the same professor in the same lecture hall. In my example, in framing this dialectic tension as a choice between claustrophobia or corpses I am referring to a specific example, a story told to me about a long-term expat INGO worker who, on returning to Goma after a short time away, felt a sense of claustrophobia (lower right cell) and decided to go for a swim in the lake, which of course is not done. The person in question felt fantastic at the sense of freedom in breaking away from the representation of Goma as a place of fear (much as I did on the moto)—until a corpse floated by (upper left cell).

The material reality of a corpse in the absolute space of Lake Kivu cannot be ignored. The more conceptual, relational, and lived aspects of the matrix contain depth of understanding, but

> There is a serious danger of dwelling only upon the relational and lived as if the material and absolute did not matter. Staying exclusively in the lower right part of the matrix can be just as misleading, limiting and stultifying as confining one's vision to the upper left. (Harvey 2006: 147)

In refusing to live in a claustrophobic state, I walked by myself from place to place in Goma and I took moto taxis at night. This is not done, especially by Muzungu[4] women. I was lucky not to have any unfortunate incidents take place; others weren't. I was told of instances when people had chosen to do the same and had been beaten and robbed; the material realities of poverty do not go away because you choose to act differently. Neither do the effects of conflict. At one point, after I had moved to Caritas Guesthouse, I was followed for quite a distance down the main road. When I turned and confronted the young man, I could see that he was disturbed and/or drugged, and given his age, thin, muscular appearance, style of clothes, and demeanor I felt the likelihood of his being an ex-combatant was high.

I don't have a solution for this tension, but the matrix helps us think about our choices—do we choose claustrophobia and its likely effects, or do we choose confrontation and its likely effects? And how much does the powerful discourse of securitization influence your choice? Some scholars argue for immersion in the community as an alternative safety strategy instead of bunkerization (Andersson 2020; Duffield 2014). Bearing in mind that this is just one anecdote, I can nevertheless finish the story of the young man following me with a demonstration of the immersion tactic. I was eventually able to shake my follower by getting into a share-taxi van. When this man tried to follow I asked the conductor, in my limited Swahili, not to let him on—immediately I was helped out.

Sending a Letter

Most expats and the middle to upper class in Goma use courier services for post. The act of using courier companies exists as an absolute experience but also as a relative experience when one considers the different space-time compression between one letter arriving and another. I tried to send a postcard via the local post office. It was a shambolic experience, not due to the customer service (the postmaster was hugely impressive) but due to the facilities. In fact, for whatever reason, the postcard never made it to Australia. The state of governance in the DRC has not received as much attention in international press as it should. Perhaps one of the reasons it hasn't is because our relative affluence (the ability to use couriers is just one example) eases our passage through these places, depriving us of the frustrations that would normally be felt by having to deal with broken bureaucracy.

The role of frustration as an emotion leading to cognitive decisions (the choice of which story to focus on) leads us back to the effect of affect, those sensate tendrils that stick upon our bodies and act upon us . . .

Dateline: Goma, DRC, August 26, 2014

I can see my young friend, my favorite smiling waiter, is struggling under the weight of the bucket of water and yet the force of that weight takes me by surprise. My arms are yanked downward, I feel my core muscles struggling to engage, and my back starts to ache. I instantly regret offering to take the bucket down to my room myself, and for a moment I am hopeful that young Judson, who next year will be at university in Kinshasa, will see my suffering and take the bucket back. But no, he only smiles more broadly, glad to be shot of it. And, I think, glad to see that I've understood how difficult a task it is.

Now the struggle back to my room, with the centrifugal force of the bucket sometimes propelling me forward and sometimes dragging me back, has become a matter of pride. In my western arrogance and habits of wastage I used up all my water and so now I must pay the price of asking for more. As a form of penance it is small and ludicrous, but I am glad to have more water.

Maybe a week later I am reminded of that arm-aching moment. I am in the Au Bon Pain boulangerie cafe after I have made the switch to the Caritas Guesthouse, run by the Congolese arm of an INGO and filled with INGO workers, academic researchers, the odd lawyer from Kinshasa, and security consultants. By the by, my room at Caritas, where there is never a lack of running water, is less clean than my room at Hotel Versailles, where there was almost never running water; the more interesting point of course being why the expat neighborhood always has running water . . . Anyway, right now, I am considering Eduard,[5] sitting opposite me, in Goma's brand-new boulangerie. The baked goods and coffee are excellent. On weekends it is filled, not just with residents of Goma, but also people from Bukavu who have made the journey, which takes between two and a half and five hours, just for the boulangerie. The trip from Bukavu to Goma can be fraught with danger. There are different militias who hold sway on the 220 kilometers of road between the two major cities of this part of the DRC, and the public passenger ferries on the lake are known to have capsized. However, luckily for those seeking European-style coffee and croissants, the major hotels in Goma and Bukavu offer a private boat option that also halves the travel time.

This is not a weekend. This is the middle of a workday, but I needed a break and, luckily, Eduard and Zara[6] walked in, Eduard and Zara, whom I first met at that trivia night. Somehow Zara and I, over coffee, tea and pain au chocolat, got on to the topic of the action TV series *Spartacus Blood and Sand*. Zara is lamenting the decline of the British male. Zara is an academic par excellence; she seems to live and breathe her profession, and in the evenings we talk for hours without pause of her day's interviews about, and with, rebel groups. But sometimes, in the middle of the day she needs a cup of tea and a safe haven to recharge her batteries and plan the next interview, and that is when she heads to Au Bon Pain, where I almost invariably am typing out stories, or sometimes pretending to do so while I listen in on conversations.

Eduard is a "fixer" who works with all sorts of international interveners. He has a wealth of contacts from his time working in the UN, first in the DRC and then in Haiti, before returning to Goma. He arranges Zara's interviews, takes her there, translates, and gives her context. He is her DRC colleague, although he will never be recognized as such in the literature (in this case it is his own decision—it can be dangerous to put your name to things in the DRC). Today our conversation is about less serious matters. We are discussing the penchant for skinny jeans in the UK and this trend in male fashion in the west in general. Among other points it has been linked to the fact the producers of *Spartacus Blood and Sand* turned to Australia to find cast members who could live up to the Hollywood image of Roman soldiers as being tall and broad shouldered, with bulging thigh muscles. Eduard is confused. "Why Australia?" he asks, and in a confused way I try to explain "the rugged outback lifestyle." I speak vaguely and in stereotypes of things I really know nothing about.

Eduard doesn't buy it. He tells me about his childhood and adolescence on the farm, carrying water, digging, working hard. Why don't I look like that, he asks. I'm not a nutritionist or a geneticist or any sort of expert qualified to respond, and yet of course in the entertaining flow of the discussion I assume a position. I argue the difference is to do with nutrition—after all aren't a bunch of the people in this café supposed to be dealing with Congolese hunger?

But for a split second, before this ridiculous, rambling conversation moves on to "nutrition," I feel the force of his argument on a different level, like the centrifugal force of the weight of that bucket of water operating somewhere in the pit of my stomach. I even fancy my upper arms sore again and in the back of my mind a second train of thought takes hold. Why do I associate health-giving hard labor with Australian farms, an image of a noble lifestyle, and almost the exact opposite, degradation and ill-health, with Congolese? Somehow it all feels tied up, Eduard's pride in the youthful hard labor of a lifestyle a world away from his own now and from his son's. My pride at managing to get the bucket to my room, the pride of the main cleaner at Hotel Versailles who looked gorgeous in his Sunday best but on other days stank with sweat and managed to keep everything spotless.

Hard Work.

Why do we undervalue the hard work of the Congolese? Part of the reason is to do with popular images, discourse, and power, as demonstrated by my attempt at a description of the mythic Australian outback. Through the distorting power of the development discourse, Congolese farm work is associated with hardship, not hard work—and yet it is both. The difference is that the hard work element is obliterated and the hardship element strengthened by the victim image so prevalent in our conceptions of sub-Saharan Africa, as equally mistaken and stereotypical as the image of the outback. There is plenty of poverty in the Australian bush, kids who come to school hungry and scabby, especially during drought. I saw that in my time as a rural reporter. There is also plenty of food in the DRC—something that was specifically pointed out to me by Aristotle[7] when he invited me to sit down with his family for dinner. I conveniently forgot both these points sitting in the cocoon of Au Bon Pain and trying to make a coherent "argument" on the spot. Could part of the reason for my forgetfulness, and the easy acceptance of the victimhood stereotype of sub-Saharan Africa, be because we don't feel the work going on around us? I mean really *feel* it, feel it in our sinews and touch it in the calluses of our hands? Instead, we only feel aggravation when things are slower or more uncomfortable than we're used to, despite the simulacra of the west around us. Or perhaps we feel aggravated with ourselves for so easily occupying these spaces of white privilege, so quickly transforming into people who unreasonably complain about their hotel amenities. We become the character we wouldn't like very much if we were reading a colonial-era novel set in this place.

If, as Freud posited, there are indeed sensate tendrils between affect and conscious thought, how much difference can carrying a bucket of water make?

Dateline: Sydney, September 19, 2014

I've stopped at a street corner. I'm leaning against the wall of a 7-Eleven convenience store and I'm scribbling in my research diary.

I feel ill, sick to my stomach.

I'm heading to a cocktail bar to meet up with a friend to celebrate my return to Sydney. It's just occurred to me that for one cocktail I will willingly hand over $20. Double the amount I would pay my translator in Goma per hour, and if someone could speak English, even if their English was not very good, I would choose not to use my translator because I fancied myself to be on a strict budget.

For many months into the future when I see a $20 cocktail on a menu I will flinch, reminded of the inequality of the world and how little value is ascribed to local languages. But, with the continued habit of ordering and paying, a callus will develop, and then the feeling will start to fade and appear only very rarely—if at all. Bourdieu is right; the social world doesn't work in terms of consciousness, it works in terms of practices.

The Space-Time of Goma's Aid/Peaceland

> Abusing my power. Full of resentment. Resentment that turned into
> a deep depression. I found myself screaming in the hotel room.
> —"Alright," an anthem of the #BlackLivesMatter protests in the
> United States (Lamar 2015: @03m25s)

Nairobi is a good place to be an international correspondent. There
are regular flights to the nearest genocide, and there are green lawns,
tennis courts, good fawning service. You can get pork belly, and you
can hire an OK pastry chef called Elijah (surname forgotten) to work
in your kitchen for $300 a month.

—Wainaina 2012: online

The decision to pay attention to the critical literature on Aid/Peacelands was
prompted by my experiences and feelings "in the field." However, as emphasized by Nash in considering the interdisciplinary nature of journalism
(2016: 118) it is for all reflective foreign correspondents to take note of the
critical literature of development as a matter of course, because development
actors are powerful players in the lives of people in borderlands and need to
be taken into account by journalists in their fourth estate role.[8] My analysis also follows what has been termed the affective turn in critical theory
(Kim and Bianco 2007). Consider the fact that Binyavanga Wainaina's 2005
satirical essay, "How to Write about Africa," talked exclusively about representation; that piece, published in *Granta*, begins simply with a list of
words. In 2012 however, his analytic focus has changed. He begins (see
the epigraph above) with a description of practices around the creation of
words and he pays specific attention to foreign correspondents, "calling
out" their lifestyle in Nairobi. Why should Wainaina (and we) care about
correspondents' lifestyles? The answer may be found in the work of theorists
of space and time who, following Marx, point out that "in transforming our
environment we transform ourselves" (Harvey 2006: 88). Braudel called the
structures of everyday life "the limits of the possible" (1981). So what are
the limits of the possible when you report on the poorest, most marginal-
ized, people in the world yet live in a world apart? How does the practice of
enjoying cheap cocktails by the lake every Wednesday in Goma (and $20
cocktails in Sydney) impact your identity as a foreign correspondent? How
does it become part of your habitus?[9] Harvey, quoting Nancy Munn, tells
us that "socio-cultural practices 'do not simply go on in or through time
and space,' but they also 'constitute (create) the spacetime . . . in which they
go on.'" Actors are, therefore, "concretely producing their own spacetime"
(1996: 215).

I have used this concept of space-time previously. Writing articles about
young Lebanese Australian leaders in southwestern Sydney, I consciously
described a local street with a traditional old-fashioned Lebanese sweet

shop. This was a representation of space (local/Lebanese Australian) and time (old-fashioned, the past). I also described them flicking between the channels MTV and Al Jazeera at a time when most Australians would not have heard of the latter channel. Thus, readers could see the social practices of my interviewees, creating a sense of space-time differing from that of the majority of the Australian public at that time—a more multilayered time and an expanded space.

In another article, dealing with the forced removal of an elderly Indigenous Australian from his home due to a controversial state government policy, I consciously sought to change the space-time associated with his place of residence. For the majority of our newspaper readers, the space associated with the words "Gordon Estate" would be considered threatening and political, and the time association would be the recurring riots on New Year's Eve that had punctuated the previous three years. In this particular article, I strove to describe a different space-time, the place of a safe family home and the time of many decades, dating right back to this gentleman's original forced removal from his home on the local riverbank. Terms like "space-time" may seem highly theoretical for the rough and tumble of a regional daily newspaper, but they're not. They consist of, simply, the conscious, reflexive application of journalistic writing tools already in use for something beyond story "color." That article played an important role in an ultimately successful campaign to grant Uncle John Hill an exemption from the forced relocation.

If I have turned this analytical lens of space-time on my journalism subjects, is it not right that I turn it on myself? In analyzing my experiences in Goma with the help of the matrix, and combining this with development literature, a particular space-time can be discerned, a space that is international *and* parochial, and a time that is ahistorical *and* nostalgic. In the descriptions that follow I will justify these particular characterizations of space and time through alignment with the critical literature on Aidlands and I will consider the implications for foreign correspondents.

Space: International

Jeffrey Jackson, in his in-depth ethnography of development workers, argues that the "first thing" you have to realize is that they live in an "international space" and starts his analysis of how "daily lives and activities make the larger processes of globalization possible" by describing his ethno-graphic research subjects in the international space of an airport (2005: 52). Jackson argues that the international space in which development workers exist nurtures their capacities as "promoters of global agendas and builders of transnational institutions of governance" (2005: xii). Wright similarly describes how the growing political influence of INGOs has been "power-fully shaped by their ability to manoeuvre effectively within and between the transnational centres of power which proliferated after the end of the cold war" (2018: 37). But the process doesn't stop in centers of power. The "travelling orthodoxies" (Mosse 2011b: 7) of development and aid workers move with them and define as international the particular corridors of wherever they happen to be.

This process of internationalization works through the valuing of the in-ternational worker over the local worker concretely, through differing sal-aries; constant connection with an international world via social media; and discussions of postings around the world, with those postings linked to the constant of career progression. And it comes through the knowledge that freelance foreign correspondents are able to fund a trip thanks to an INGO based in the west, or perhaps through renting out their apartment in New York. The space is also international in its "whiteness," where whiteness is not a color but a relationship that gives the international INGO worker (es-pecially if they actually are white) greater access to power (Kothari 2006) and allows easy entrée to the world of experts and professionals for ad hoc visitors like me who just rock up.

This international space is further created through *accountability links*, whether for aid programs or articles written, which do not flow into the local community but instead head "back" to the "center" in New York, London, or Sydney. This "upward" accountability process, in regards to development, including its negative side-effects, has been described all too clearly by Easterly (2006). Important recent work by journalism academic Wright (2018) has found parallel upward accountability dynamics in place in media production processes, with African NGOs liaising with locals but subordinate to INGO media teams or multinational corporate social re-sponsibility teams. Furthermore, freelancers, correspondents, and editors may have had a strong belief in the value of news reports from Africa, but they were still ultimately influenced by news headquarter concerns such as brand positioning, awards, and fulfilling statutory broadcasting responsibilities.

Among the impacts of living in this "international" space is that it engenders segregation from the local. Former head of country office in Bolivia for the UK Department for International Development Rosalind Eyben has described the resistance of her staff, and other development professionals, to the idea of taking regular local "reality checks" in the community:

> To understand why the local may be irrelevant to donor staff even when living and working in an aid-recipient country, it is helpful to distinguish between place and space. . . . The Internet revolution has enabled the local place and the space of engagement connected with it to become increasingly irrelevant for those with global agendas. Communication is no longer dependent on geographical proximity but on other kinds of proximities— linguistic, political, cultural, ideological, etc. However, some sense of local place and community is still necessary to support a life in a global space. Most people cannot live in an entirely virtual world, and this is why sociality is so important in the everyday life of donors. Hence, the intense social round of meetings, lunches, receptions and picnics were a continuous effort to recreate and sustain the global policy space and to prevent the local political space from disrupting it. (Eyben 2011: 155–56)

Thus, by international, we also mean, specifically, *not* local. The boulangerie typifies the international connection and local segregation. Like an expat moth to flame I discovered it on my very first foray into the city. Expats from other cities in both the DRC and neighboring countries knew of this "talk of the town"—yet many local moto drivers hadn't heard of it. The boulangerie was obviously for internationals and elites—not the hoi polloi. At times it resembled nothing so much as a packed coworking hot-desk space. Strangers would share tables, tapping away on their laptops for hours, and of course they weren't strangers by the end of the visit. People would come and go, and greetings and introductions were manifold. These interactions were based not just on the particular space of Goma but connected to the international world of Aid/Peacelands. Several of my research interviews resulted from introductions made at the boulangerie, and three of my research interviews took place there. At one point an interviewee, despite knowing me to be a journalist and researcher, asked if I was part of the "UN family" ("Interview F" 2014), clearly implying a distinct international and mobile tribe, mobile in terms of both moving between locations and moving between strands of international knowledge-work.

Scholars Rajat and Stirrat argue: "In general there is relatively little social contact between development professionals and the host population at the level of friendship" (2011: 168) except for a local subset of the international tribe: "the "comprador" class [government personnel, employees of INGOs, servants, fixers], intermediaries between the host population and the foreign development workers who control communication and usually produce what is expected of them" (2011: 168). In other words, the "comprador class" skew the world to suit the expats' understanding of it, which again reinforces the space as international.

What does this mean for foreign correspondence? When working directly with INGOs Wright found that there was little opportunity for inclusivity or empowerment for interviewees, even when those interviewees were found through local NGO partners. Wright (2018: 257) found "field workers' own economic interest and their previous experience of donor reporting" could "lead to them framing media participation to local people which limited what they [the locals] could say."

Outside of controlling the interactions around interviews this comprador class obscures the local in quite subtle ways. For example, a journalist operating with a fixer in Goma will lose any sense of the broad middle reality for the Congolese. Fixers find hotels where the electricity and water works and they pay/talk their way through police roadblocks with minimal delays. And, at the other end of the scale, they will locate tragic victims of war. The argument that international journalists need to capture and understand everyday connections, and they cannot do so if they are not living them, has been made by foreign correspondent Anjan Sundaram (2014, 2016). Sundaram states that embeddedness in everyday life is necessary for the proper interpretation of events. A related argument is made by Ben Rawlence, author of *Radio Congo*, who says that to create new knowledge of a place like the DRC, "you need to get the balance right between reality, which is horrific, and the encompassing reality, which is human and normal and mundane" (Rawlence in Taylor 2012: online). This is not to suggest that foreign correspondents become locals; on the contrary, they have to inhabit an international sensibility to craft news reports aimed at fostering a sense of the global public sphere. However, this sense of a global public sphere is also jeopardized if it is an exclusively international habitation resulting in the erasure of the broad multilayered local reality where so much of human endeavor resides.

Time: Ahistorical

The lacuna effect identified in chapter 1 introduced us to the imagined ahistorical nature of both Africa and the development promise. So it should come as no surprise that Apthorpe characterizes Aidland as "*History nullius*" (2011: 210). The sense of time is altered because the category of the past is conveniently forgotten, and also because of a lack of continuity fostered through short-term postings. Rajak and Stirrat point out that there are "relatively few people in the development industry who have permanent contracts" and even those are cycled through two-year postings abroad (Rajak and Stirrat 2011: 164).

Ahistorical characteristically means apolitical, with apolitical including the inability to see history in the making. For example, Eyben tells of a meeting at the Ministry of Finance in La Paz where there was rioting on the street outside the ministry: "Those of us inside in a meeting with the Director General and his staff were wiping our eyes from the tear gas that had seeped through the cracks even as we continued to ignore the riot and discuss the contents of the Poverty Reduction Strategy" (2011: 154). It is inconceivable that a journalist in that situation would have ignored the riot. However, it is not only conceivable but likely that foreign correspondents would turn to these international interveners for interpretation and analysis of the event they were assiduously ignoring. Eyben goes on to note:

> The Bolivian officials and consultants who came to our *receptions, dinners* and *picnics* had a shared understanding with the donors of the way the world worked. This understanding became so deeply ingrained that *we were no longer capable of imagining other approaches* even when the views of locals broke rudely into our deliberations. (2011: 154; my italics)

Thus we see Eyben pointing out the relationship between apolitical/ahistorical time and international sense through closed social networks. There is also a relationship between the ahistorical and the parochial.

Space: Parochial

Parochial space-time is produced through the effects of ahistorical space-time. Harper (2011) details that when international development workers

go to Nepal they ignore the history of indigenous Nepalese health services. This in turn shunts them toward parochialism by blocking their ability
to be cosmopolitan, with cosmopolitanism defined as "that [which] draws
on a sense of universal humanity, reflexive distance from one's own culture and relating to traditions other than one's own" (Anderson in Harper
2011: 124).

The parochialism of Aidland is described as a dislocated tribe "socially at
least as restricted as any other strong ethnic identity" (Friedman in Mosse
2011b: 14), a "monoculture" (Eyben 2011: 152) and a "ghettoized existence.
Development workers tend to exist in a social cocoon, socializing with each
other and reproducing not only differences between expatriates and the host
community but also national differences amongst the expatriates" (Rajak
and Stirrat 2011: 169). I experienced this contradistinction between expat
communities myself in Goma. My second Friday night there (the first Friday
night was the night I arrived), I went to the local nightclub frequented by
expats. I was introduced to a senior MONUSCO official, an Australian, who
berated my Congolese friend for not bringing me, as an Australian, to him
immediately on my arrival. In an "ocker" Australian accent much broader
than I normally hear in Australia, I was told not to worry—if anything
happened, he would make sure I got out of Goma safely. I was embarrassed by
this nationality-based favoritism but also extremely pleased to have his card
in my pocket. Yet, properly analyzed, that action of pocketing the card stands
for everything I stand against. It reinforced my parochial Australian sensibility in a place where my professional role rests on the ability to transmit a
global sensibility. Furthermore, this parochial nationalism, practiced by all
groups, strengthens "solidarity among whites" across all nationalities; it is not
only colonial in its origins, it is classist. Solidarity among whites in colonial
and postcolonial contexts "in its curious trans-state character . . . reminds
one instantly of the class solidarity of Europe's nineteenth-century aristoc-
racies, mediated through each other's hunting lodges, spas, and ballrooms"
(Anderson 1983: 139). Is there really any difference between a ballroom and
a nightclub?

Time: Nostalgic

I have adopted the term "nostalgia" consciously as being seeped in sentimentality, for sentimentality disregards the power of history—it is again

ahistorical—for only ahistoricism could explain the colonial nostalgia evident in the Aidlands of the postcolonial era. Rajak and Stirrat (2011) describe the nostalgia of Aidland as a yearning for a "simpler" time world, "either a nostalgia for an imagined past of home or an 'imperial nostalgia' for a world which notionally preceded the discontinuities, disjunctions and displacement of the world of development" (2011: 174). However, the point I wish to highlight is that the simplicity of this colonial world was built on clear demarcations, racial demarcations.

In the self-contained international, ahistorical, parochial time-space of Aidland, this major truth is conveniently forgotten and, instead, the only politics that come into play are self-referential to the Aidland community, for example, over the choice of hotel:

> Where visiting consultants should stay is often a complex issue which involves consideration of status, facilities and cost, but also consideration of style. Frequently colonial style hotels are favoured over modern counterparts. Indeed snide comments are often made over which people from which agencies choose which hotels. "Old colonial" is somehow morally superior to "new capitalist." (Rajak and Stirrat 2011: 172)

Of these four elements of space-time, described in the Aidlands literature, the nostalgia element, especially in terms of colonial nostalgia, did not seem very strong in Goma. However, I have chosen to include it in this general description of the space-time of Aid/Peacelands, both because of the strong arguments made by Rajak and Stirrat and because I have witnessed this colonial nostalgia elsewhere in Africa, and indeed in the old colonial capitals of London and Brussels. This nostalgia may in fact be triggered by colonial-era architecture, and there is not much of that in Goma, except perhaps the manicured, formal gardens, maintained by gardeners with hand shears. As far as I could tell, there was also not a lot of nostalgia for "an idealised 'back home'" where things work properly (Eyben 2011: 148), although the excitement over the quality bread and coffee in the new boulangerie does testify to at least some of this feeling, even if espoused via excitement rather than complaints.

These caveats reinforce the importance of situational specificity. Goma is a city that only hit the international radar in 1994. Given its relative insignificance during colonial times, and its deep intertwining with the modern form of ongoing warfare, I would like to tentatively propose that a different sort of yearning for simpler times is visible—a nostalgia for the neoliberal promise of freedom and for the early days of UN "success," when young westerners felt they were truly and clearly "helping" and celebrated that success with gusto. See, for example, the account of the Cambodian elections in *Emergency Sex* (Cain, Postlewait, and Thomson 2006).

The youthful "party time" nostalgia visible in the intense partying and the constant movie, music, and pizza nights is spatially connected to the

> fortified aid compound [which] is more than a defensive structure; it is a therapeutic refuge that both separates international aid managers from outside uncertainty, and encloses the supportive social networks and cultural props that allow for narcissistic forms of care-of-the-self. (Duffield 2013: 58)

In existing inside the compound walls, or in only certain sanctioned spaces, this party-time nostalgia is also linked to the biopolitics of securitization, and it is important to recognize this connection because of the sensate tendrils between affect and consciousness. For example, Duffield argues that bunkerization creates a "postpolitical" reality, and that in "seeking refuge from uncertainty, today's social and economic elites are unlikely to imagine Utopias that involve a better future for all. Futures are more likely to be imagined in terms of the exclusivity that bunkered-life vainly promises" (2013: 58).

It is not necessarily a journalist's job to "imagine a better future," but journalism, existing in, and justified by, the concept of the public sphere is always political and must feel itself so. Relatedly, it is a journalist's job, and particularly a foreign correspondent's job, to connect to all people; exclusivity has no role in journalism.

Colonial nostalgia, anywhere in the world, cannot help but disconnect the foreign correspondent from a proper political understanding of the postcolonial society of today—the society that has become. Likewise, in Goma, hedonistic and securitized party-time nostalgia is too prevalent and structured to be downplayed as simply the natural human emotion of having fun. Party-time nostalgia reinforces neoliberal structures of individualized action (it's no coincidence that "work hard / party hard" is also the motto of global financiers) and helps occlude hierarchies—in both work and play—in which we are complicit.

Cultivating the Doxa of Globalizers

Who is this person? Sitting in this bar, working in this café, dancing/
networking in this club. Who is she? Oh fuck, she is me.
—Post-it on my wall describing my time in Goma

The intricacy of social control by spatial ordering (and, conversely,
the complex ways in which social orders get challenged by the transgression of spatial boundaries) requires sophisticated analysis.
—Harvey 1996: 230

As described in the literature on embedding, and as experienced by myself in Goma, aid workers and journalists share the same world of hotels,
chauffeured four-wheel drives, relief flights, restaurants, and bars. They
also share similar cultural backgrounds and motivations (Chandler 2013).
Kate Wright (2018) has shown how the world of INGO staff and freelancers,
and British journalism staff and freelancers, working on African stories, is
small, with key overlaps and movement between professions. As well as
sharing social contacts and cultural similarities, many share similar normative concerns. Her work brilliantly reveals how this leads to an overlap
of agendas, including "inter-elite insularity" (2018: 147) that was "rarely
reflected upon" (2018: 265). In a case study of reporting from the DRC,
Wright discusses how interaction over a particular story facilitated "normative and organizational changes already underway at both the INGO
and news outlet: with the former remodelling itself as a kind of news wire
agency, while journalists at the latter reconstructed public-service journalism as including 'witnessing' and 'human rights reporting'" (Wright
2018: 148). My argument is that Aid/Peaceland space-time is another element that cannot be ignored when considering this growing entanglement of professions. A useful parallel can be drawn with Australia's capital,
Canberra, an isolated city created for the sole purpose of governing, where
journalists and politicians work and socialize together, sharing the same
spaces and existing in a career-focused high-stakes hothouse environment
away from their families. In her controversial exposé describing the inner
workings of the Canberra press gallery, Margaret Simons writes: "It is a
place where if you refuse to blend the lines between the personal and the
professional, then your effectiveness is reduced" (1999: 50).

The Canberra press gallery, Westminster lobby, and Washington's Capitol Hill are recognized as places of power, and their dynamics are analyzed and discussed in both academia and popular culture. Those discussions include concerns that a shared social space might lead to a shared social view of the world at large. In her book Simons discusses the various public debates during her time of writing, including the view "that journalists might broaden their minds if they were not living cheek by jowl with the politicians" (Grattan in Simons 1999: 79–80).

The hotels, cafes, clubs, and restaurants of Goma, the spaces of Aid/ Peaceland, should also be recognized as scenes of power that enable limited frameworks of interpretation. What's more this argument about the role of space-time in determining frameworks of interpretation can be developed further with the help of Bourdieu's theory of *doxa*. Bourdieu contends that all fields (and subfields) operate through a sense of *illusio* and doxa. These two concepts, taken together, can best be understood as the "logic" of the field. *Illusio* can be understood simply as the idea held by any agent in the field that the game is worth playing, and doxa can be understood as the "rules of the game," a shared "universe of tacit presuppositions that organize action in the field" (Neveu and Benson 2005: 3). Thus, when Wright (2018: 146– 47) describes how both journalists at Channel 4 News, and media workers at Human Rights Watch, judge their reportage as successful by putting pressure on specific Department for International Development policy actors, we see a shared *illusio* that this elite game of international humanitarian governance is worth playing even though it may come at the expense of focusing on other, more demotic, forms of journalism (Wright 2018: 261–62).

To return to doxa, it is important to understand that "the rules of the game" are not written down anywhere, nor is doxa encapsulated in the presuppositions that organize action. Doxa is also those actions themselves. Precisely because there is no rulebook, it is practices that help legitimize the dominant ideas, or orthodoxy, of the field. Bourdieu says that doxa operates as a form of domination "through language, through the body, through attitudes toward things that are below the level of consciousness" and that "the main mechanism of domination operates through the unconscious manipulation of the body" (Bourdieu and Eagleton 1992: 115). It is in this description of doxa where we can see direct links to the role of space-times and of affects.

A clear example of the power of space-time on doxa can be seen in a specific case where the space of foreign correspondence was consciously manipulated by a politician. Former US secretary of state Henry Kissinger chose to try to change the "rules of the game" by holding Middle East peace talks in Jerusalem instead of Tel Aviv. Lederman writes that this "altered [the] perception of what the Middle East problem was all about" (1992: 100).

> Just being in Jerusalem for extended periods had a profound symbolic impact on the foreign journalists. Tel Aviv [where the foreign correspondents were normally stationed] was the seat of the Defence Ministry and the IDF's General Staff, and its very proximity to where the journalists lived and worked set much of the tone for their coverage. Jerusalem was the seat of government and the center for international, if not domestic, politics. The very act of being in Jerusalem and reporting from the city could not but change the focus of reporting.
>
> When journalists exited the Government Press Office in Tel Aviv, they faced the Defence Ministry-General Staff complex just up the street. When journalists looked out of the windows of the King David Hotel in Jerusalem, where Kissinger stayed, they saw the spotlighted walls of the Old City and the Arab sector of town. (Lederman 1992: 100)

This example eloquently highlights the importance of space-time as an active agent in the field—it highlights why the Aid/Peaceland space must be engaged with critically. The change in foreign correspondence brought about through Kissinger highlights another of Bourdieu's arguments, that the field of journalism has "very low autonomy" as a subfield with the field of power (Bourdieu 2005b: 41). This low autonomy means that the doxa of our field is constantly shaped by external pressures, for fields are never static: "A field is a field of forces and a field of struggles in which the stake is the power to transform the field of forces" (Bourdieu 2005: 44). This field will be subject to internal pressures and external pressures, but fields with low autonomy (e.g., journalism) are impacted more by external factors than fields with high autonomy (e.g., art). Because of our reliance on the field of politics, Kissinger was able to dictate the space where journalistic interaction would happen.

If fields are a site of struggle and doxa is constantly being reviewed and reformed through practices, affects, and spaces, then the international, ahistorical, parochial, nostalgic space-time is doxa in creation, a doxa that states the rules of the game are neocolonial. In the final journal article of his life, published posthumously, Bourdieu described a new international economic field visible through the politics of structural adjustment, which disproportionately disadvantaged African economies (2001). Bourdieu wrote that, within this field, there are subfields at work, and although Bourdieu didn't elaborate on these subfields in that article, the field of aid-development-peacebuilding economics and politics must be one of them. Indeed, this has been argued by Jackson, who, expanding on Bourdieu's article, identifies both journalists and development/aid workers as "globalizers" (2005: 3).

I am not here suggesting that the doxa and *illusio* of foreign correspondence can be reduced to the pattern of relations in Aid/Peacelands; journalism is a subfield in itself with its own rules. However, like other scholars (Franks 2013; Harris and Williams 2018: 69–72; Powers 2018; Wright 2018), I am highlighting the interdependence and porous boundaries between the fields of INGO and journalism work. More especially I am seeking to start a discussion on how this overlap takes place in specific Aid/Peaceland space-times that create their own dynamics and that are hindering the process of decolonization. To decolonize foreign correspondence from problematic aspects of this doxa of globalizers, the various ways it operates must be examined. In this chapter I have focused on introducing the role of space-time as an active agent in shaping patterns of relations and patterns of thought. In the next chapter I will drill down to look at one of the clearest and most researched examples of problematic neocolonial practices in borderlands, the relationship between foreign correspondent and fixer.

4

The Hero Correspondent and
the Hidden Fixer

If white, western journalists are more worthy of receiving both pro-
active and reactive assistance with safety, then it appears that the
mainstream, English-language war reporting industry does not
ethically recognize the value—indeed, the humanity—of its locally
based employees

—Palmer 2018: 104

Let me begin this chapter by telling you a story—a story I feel I need to tell—
and yet the very act of telling that story is something I wish to critique.

In 2011 it's quite possible that a camera strap breaking saved my life.

This was during the visit to the Kosti refugee camp in Sudan.[1] Aware of
some of the critiques I have developed in the previous chapters, and particularly wanting to avoid the white angel of mercy frame in our reporting,
my partner and I had extremely limited contact with INGOs and did not organize permission to enter the camp but simply hopped on the local bus and
walked in. Thanks to a high level of English skills in Sudan, we were able to
talk to most people we encountered. One interview I conducted was with
three teenage girls hanging out together listening to a radio, hiding from the
sun by lying under a camp bed. Their story seemed important to me because
these girls had been born in the north and so were heading to a "homeland"
they had never seen. The three had varied and conflicted feelings about the
move, but they were now stateless and had no choice. I considered it important that they had missed out on several months of schooling while waiting—
and they themselves were very much aware of their interrupted schooling.
I wanted to tell their story and I wanted to illustrate it with a picture of them
under the bed—highlighting the lengths they had to go to to find shade.

Borderland. Chrisanthi Giotis, Oxford University Press. © Oxford University Press 2022.
DOI: 10.1093/oso/9780197565797.003.0005

As soon as I asked to take a picture, one of the three girls refused to be photographed and left. If I had been smarter I would have taken more no-tice of her reaction. Instead I photographed the remaining two girls again and again, determined to get the perfect shot and struggling with the con-trast between the midday sun and the shade under the bed. This was my first mistake for the day. Until that point, we had been moving fairly quickly and inconspicuously through the camp; now a large crowd had time to gather around us, and among them was at least one person with an agenda. To cut a long story short, when we boarded a public bus to go back to town, it was surrounded by police and we were told to get off. I was asked for the camera, which I refused to give, not wanting to lose it or the precious pictures. Finally, as the number of security personnel around us, and the variety of their uniforms, increased, I offered to delete the photos from the camp. This was my second mistake. Deleting the photos would have removed the evidence that the informant was relying on to get his kudos for capturing two "western pornographers" who had been photographing girls under beds. He snatched at the camera, and I instinctively fought back. I didn't know it, but one of the uniformed men had raised the butt of his rifle ready to hit me in the head. The camera strap broke, the informant was left with the camera in his hand, the situation was defused and I never got to experience exactly what damage that blow to the head would have wreaked.

We were bundled into an army utility, complete with a machine gun welded onto the tray, and taken to a security compound. We were told we would be imprisoned for seven years. Luckily, we managed to convince the commander who interviewed us that we were not journalists (or pornographers), just dumb tourists who had come to Kosti because of its famous fish market. Did he really believe that? I suspect he just didn't want the trouble of arresting us. To quote from Steve's blog describing the arrest: "The camera was returned in one piece, less every photo taken that day. . . . Deleting all the photos proved their 'pornography' angle was simply a ruse. We were detained because the authorities wanted to cover up what was happening at the 'returnee' camp in Kosti" (Madgwick 2011: online).

I have told this story for interrelated and conflicting reasons. First, it is a demonstration of the very real danger journalists face when they enter refugee camps without official permission gained via the help of INGOs or fixers. Yet we couldn't have entered with an INGO because they were not publicizing the returnee situation at the time. Thus the story also highlights the distance between the two professions of aid and journalism, and the point that journalism is always political in its nature. It is unimaginable that journalists would have sat in a ministry office in Bolivia ignoring tear gas and rioting as described by Eyben in Chapter 3 and neither could Steve and I imagine ignoring the situation in front of us in Kosti New Port. This *inability not to report—an inability not to add to the debate in the public sphere—* is understood by the public and by authorities, and it is part of the reason why journalism is *always* political—the mere presence of a journalist shifts the situation into the public sphere—and this in itself adds to the danger experienced by international correspondents *and* more especially by local journalists and fixers.

I also told that story because I am afraid—not out in the field with soldiers and guns but here, writing these words—I fear the reaction of other journalists when reading this book and, in particular, that they will classify me as an academic sniping from the safety of my computer screen. It is important for my sense of having a "right" to talk about these issues that it is understood that I know the dangers faced by members of this profession. And I am not minimizing the impact of those dangers, on individuals and on the overall environment of press freedom. Yet, ironically, this *need I feel* to talk about my "credentials" in foreign correspondence is exactly what I wish to critique. I remember one meeting with colleagues in 2012, when I had returned to Sydney and become a tutor in the journalism program at the University of Technology Sydney. Somehow, the conversation got to a point where I said the words "when I was arrested in Sudan." We were all journalists and Bourdieusians, and one of my colleagues commented, "There is Chrisanthi demonstrating her cultural capital." But why? Why should my cultural capital be that I was *arrested* in Sudan? What does the fact that I was arrested tell any of my journalism colleagues about the quality of my reporting? About how well, or how badly, I told the complex story of the stalemate at Kosti New Port, and the situation's impact on the people stuck there? My standing as a journalist should not be based on barroom tales of bravado. The fact that a story of being arrested equals cultural capital, and more especially for white correspondents, simply shows the extent to which foreign correspondence is affected by a "hero" stereotype. It is this hero stereotype that we must take issue with if we are to make the profession fit for purpose in today's world, because the "hero" foreign correspondent is deeply entwined with a problematic colonial heritage and also obscures the current reality where the journalists bearing most of the danger are not the elite foreign correspondents.

Heroes, Plural

That foreign correspondents inspire a hero image is well documented (Murrell 2009; Tumber 2013; Ayres 2006; Leith 2004: xxii) and this image impacts decisions made by journalists in a global world full of conflict. Today, with the splintering of the news ecology, young journalists seek to trade on the cultural capital of the hero foreign correspondent for professional advancement. Describing conversations with journalism students, BBC World Affairs producer Stuart Hughes notes that often they have no idea why they want to go to a conflict zone, and what sort of stories they are interested in telling. Their interest instead comes from the perception it is "a rite of passage for a young journalist" almost like a "gap year" (cited in Cottle 2016: 135–36). Liebes and Kampf identify the TV genre as responsible for the rise of this "performance journalism": "Whereas traditional models position the social/moral/structural issue up front, for the new performer journalist the path to such an issue (if at all) leads through an individual story, built on action, drama, tragedy, or all three combined" (2009: 243). They argue that performance journalism is on the rise in all genres of journalism but "the chaotic borderless scene of fighting terrorists offers ideal conditions" (2009: 243).

Yet there is also movement in the field which seeks to decenter the correspondent. When ABC journalist Sally Sara was asked to write about her time as Africa foreign correspondent, she rejected the common form, writing: "I knew that I didn't want to write a 'foreign correspondent' book" (2007: xiii). Instead, she created a work consisting of in-depth portraits of African women. Other foreign correspondents, while not necessarily challenging the genre through their own productions, do interrogate the attitudes that feed it: "Many cocky people out there who think they are hot shit . . . [and] forget that it is an incredible privilege and a responsibility" (Lorch in Leith 2004: 233). Many reject the glamorization and the "powerful illusion that here you are, happy people having adventures all around the world" while local reporters "who cover conflict all the time because the conflict happens to be taking place where they live" are never glamorized (Goldenberg in Leith 2004: 153). Colleen Murrell, in her groundbreaking work detailing the crucial role played by fixers in the international news process, spends some time documenting the hero stereotype of "autonomous and independent loners . . . swept up by great events and overcoming them by individual cunning and skill" (2009: 7) because of the way in which this hero stereotype has helped hide the role of the fixer from the public.

The glamour attached to globetrotting adventurers is also tied up with the importance of what is known in the profession as a dateline—a fancy way of saying the location of the journalist. In the time before instant, globalized communication, the dateline meant something. To find out what was happening you first had to travel to the place. These days, too often, the dateline has become a device for maintaining an image. Joris Luyendijk, in an autobiography that covers his time as a Middle East correspondent, documents this extensively; during Operation Desert Fox in Iraq he was

> in Amman summarizing press releases from Hilversum on the Baghdad bombings instead of the person receiving the wires in the Hilversum studios doing it. "From Amman" sounded better . . . Editors-in Chief judged their correspondents and reporters by the dateline . . . that's why those grown men had cried at the gates of the Iraqi embassy in Amman [at not being able to get a visa for Iraq]. Of course, if they had been in Baghdad, they'd have been immediately confined to their rooms and condemned to using the same news agencies as I was in Amman . . . but there, at least, they would have "scored" [a better dateline]. (2009: 19–20)

Speaking to academic Denise Leith, foreign correspondent Max Stahl also critiqued the pure theater involved in datelines, identifying the time compression of 24-hour news as being partly responsible for this farce, with "reporters answering live questions to studio three times a day" (Leith 2004: 337). However, this time-wasting theater wouldn't be necessary in the first place if it weren't for the preexisting glamorous image of the foreign correspondent in the field. Armoudian's (2016: 34) more recent research with foreign correspondents also found interviewees critiquing their colleagues who were "feeding the beast in the theatre of journalism" not least because of the way this style of reporting equaled "no context" and "very little interest in what the locals had to say."

The theater of the dateline was perhaps most visible in the proliferation of reports from Middle East hotel rooftops during the second Iraq war. Even as the war coverage wound down, allowing more time to leave the base, the confinement to hotels grew worse as the danger for foreign correspondents increased. Eventually, this led to two strands of important critiques. The first critique to evolve asked why, in this age of globalization, the "white middle-class male reporter" was needed (Sambrook 2010: 47). If locals were on the ground doing the work, why not simply use local reporters for all of the coverage (Sambrook 2010: 48)? Richard Sambrook investigated this question for the Reuters Institute for the Study of Journalism, asking: "Are foreign correspondents redundant?" Sambrook concluded that, in the future, more news organizations would use local reporters instead of flying in foreign correspondents, and that while the trend may have started as one driven by economics, such news organizations, especially those servicing global audiences, now consciously valued the expertise offered by locals (2010: 49–51). Al Jazeera, for example, calls this "journalism of depth" and specifically contrasts it to "parachute journalism" (2010: 51). Sambrook also pointed to BBC internal audience research that found "having a familiar accent does increase the likelihood of audience engaging in a story." However, "If UK reporters are unavailable at the scene they are satisfied with local commentary—translated if necessary. This suggests their tolerance for local reporters may be higher than assumed" (Sambrook 2010: 64). If foreign correspondents are hindering the cosmopolitan impulses of audiences they may in fact be worse than redundant. Foreign correspondents can only justify their existence going forward if they are enhancing, rather than retarding, global sensibilities.

The second stand of critique asked why the reality of hotel journalism was not being acknowledged to the public. As Robert Fisk put it:

> Why do not more journalists report on the restrictions under which they operate? During the 2003 Anglo-American invasion, editors often insisted on prefacing journalists' dispatches from Saddam's Iraq by talking about the restrictions under which they were operating. But today, when our movements are much more circumscribed, no such "health warning" accompanies their reports. In many cases, viewers and readers are left with the impression that the journalist is free to travel around Iraq to check out the stories which he or she confidently files each day. Not so. (Fisk 2005: online)

As the prolonged and dangerous nature of borderland wars has increased, the way journalism knowledge-production practice has changed deserves to be understood by audiences. By not acknowledging the limits they are operating in, foreign correspondents not only do damage to the truth of the individual stories but also fail to give audiences a broader context to understand the place and situation they are reporting on (Luyendijk 2009: 101–8; Lederman 1992: 110).

A recent tranche of important industry and academic research has highlighted the fact that foreign and local journalists and fixers are increasingly operating under dangerous and deadly conditions and increasingly targeted for doing their jobs (Palmer 2018, 2019; Armoudian 2016; Cottle, Sambrook, and Mosdell 2016; Picard and Storm 2016; Harris and Williams 2018). This focus on press safety is an excellent development that can be used to highlight the realities of borderlands. However, there is also a trend in the practice of safety culture that is worrisome. While the literature highlights that *local* journalists are the ones facing the majority of violence and death, Palmer (2018) has identified a neocolonial practice evident in the different attitudes to safety culture. Palmer (2018) notes hazardous environment training and other protective measure are focused on Anglophone journalists, while local journalists and fixers are not only supposed to be able to take care of themselves but are also supposed to bear the responsibility for keeping the fly-in foreign correspondent out of danger. As noted in the epigraph by Palmer at the top of this chapter, this is a dehumanizing practice that values relationally white lives above others.

The Iraq war and the resultant ongoing conflict was a turning point in bringing to light the problematic practice of valuing international and local lives differently. In part, because of the extreme security risk, fixers have been converted into local stringers, or "proxy journalists" (Palmer and Fontan 2007: 6). The ethical issue here is pointed out by Palmer and Fontan in the gruesome statistic that in the first year after the 2003 invasion of Iraq, the body count for media personnel was split roughly equally between foreign and Iraqi nationals. However, as time went on, the vast majority of media professionals killed have been Iraqis (2007: 6). The practice of western journalists retreating behind secure compounds may have been initially justified by the belief that in Iraq they were more targeted than locals (Veis 2007: online). Yet, as summed up by correspondent Lindsey Hilsum, journalism is a dynamic political process: "There are occasions when we are safer and occasions when they are safer" (Cottle 2016: 139). The different attitude to safety culture highlighted by Palmer cannot be justified. It speaks to a broader journalism profession, which has yet to decolonize.

Decolonizing Relationships, Experiences, Knowledge

Lecturing in the postgraduate subject of journalism studies at the University of Technology Sydney in 2015, I was privileged with one of those rare and joyful teaching experiences. After briefly covering the history of foreign correspondents and their images as heroes, colonialists, and spies, I played, without preamble, a short clip from *Foreign Correspondent*. This was a 2013 episode that, to my mind, featured a scene that encapsulated the western journalist-local fixer relationship. In the Malian desert, the local fixer was on his knees digging out a bogged Toyota Landcruiser (that ubiquitous vehicle of the westerner in developing countries). An entourage of security and other personnel jostled for space to push the vehicle; meanwhile, the western journalist looked on and paternally clapped when the job was done. As soon as I hit pause on the video, at least two or three students spontaneously broke out with exasperated comments of "Just push the car!"

My joy in hearing the students' outbursts was based on a very simple idea: in their interactions, foreign correspondents need to stop acting like colonialists. Palmer (2019) traces the practice of employing fixers back to colonial times, and clearly, the Australian correspondent's bodily practice in that clip was not one of solidarity with the Malians, but rather one that more closely resembled the colonizer-correspondent Henry Morton Stanley being carried through the forest, in a chair, during his "achievement." Actions like these not only activate the modern culturally coded racism of different species life (the correspondent's life is more valuable, so why should he push) but also devalues the importance of locals in foreign correspondence whose roles are far more important than that of mere logistics.

Paulo Nuno Vicente quotes one of his research interviewees, a freelance journalist in Kenya, as saying it's both "patronizing" and "nonsensical" that "western media send their reporters parachuting into other countries and expect them to have the best reports" (2013: 45). It's true, it is patronizing and nonsensical—but of course that's not what happens—because those journalists then turn to the knowledgeable local fixer to help supply the story. As discussed in chapter 3, they are part of the "comprador" class and as such help perpetuate the idea of the traveling international expert. Hannerz notes:

> As in anthropology, where over the years the field workers' multipurpose local research assistants have mostly been left invisible in the resulting

ethnographies . . . , the critical importance of local helpers in foreign news work tends not to be acknowledged. (2004: 154)

Generally speaking, the flurry of industry and academic discussion over the last few years about fixers has not translated to an appreciation among audience members for their role. The 2021 US withdrawal from Afghanistan and the desperate scenes at Kabul airport helped to highlight the crucial role of local fixers, for all sorts of international professionals, through their attempts to escape the country. However, exactly what fixers do in journalism beyond translation would still be unknown. And to be fair, most audiences don't know what producers do either, or their influence on story creation. Nevertheless, there is a difference between the invisibility of a behind-the-scenes producer in a national newsroom and the invisibility of the behind-the-scenes producer for a foreign correspondent; power and race dynamics must be taken into account. In the case of a western correspondent flying in and being seen to quickly and eloquently sum up the situation, the potential subtext, for some audiences, will certainly be a variation on the theme "See, all it takes is a little western know-how and everything could be worked out."

At times this failure to acknowledge fixers has to do with protection. Palmer's interviews with fixers around the world revealed that byline attribution is a complicated issue and not just for the safety factor. Some reported byline indifference or ambivalence, simply happy to know in themselves that they had done their job helping the correspondent get the story. Nevertheless Palmer concludes byline attribution should be the regular practice unless there is a reason not to. "By refusing to regularly and systematically give credit to news fixers for their work, these industries tend to give the impression that their correspondents are the sole laborers in the field" (Palmer 2019: 190). In Goma, as part of my ethnographic analysis, I conducted six in-depth interviews with fixers, and the majority took the lack of acknowledgment as the norm, even if there was a sense of sadness about this. The exception was when it came to documentaries. Here, one fixer quite rightly pointed out that he expected and hoped to be acknowledged in the credits, for this would help him get extra work. He told me that, on one occasion, when he was not acknowledged, he asked the documentary maker and he was not given a clear explanation. The fixer then suspected that the foreign correspondent's "ego" was the reason ("Interview C" 2014).

Preparing an article on the role of Iraqi fixers/stringers/journalists in reporting for western news organizations, Greg Veis found himself rebuffed by nine of the United States' biggest news organizations. One former Iraqi reporter for a major US newspaper told Veis he felt those organizations were hiding the bravery of the Iraqi journalists so the American journalists could get the credit (Veis 2007: online).

Veis's analysis is congruent with other dimensions of the discussion so far around the impact of Aid/Peacelands in that the truth about the fixer's role interferes with the myth of the international expert—in this case the solitary foreign correspondent surviving on wits alone—and it also interferes, specifically, with the foreign correspondent hero stereotype. However, by not acknowledging fixers' existence we are giving a misleading impression to the audience that denies the complexity and specificity of the situation.

And perhaps it is not only the audience that we are deceiving—we may be engaging in a level of self-deception. The fixer shields the journalist from feeling discombobulated in the new environment, allowing the foreign correspondent to maintain their self-image of worldly professionalism. Bourdieu's emphasis on social practices (Bourdieu and Eagleton 1992) highlights the fact that one of the ways we learn and understand difference is by *feeling* it; a healthy respect for local knowledge is more likely be engendered when we feel at sea. It should be possible to look at the fixer/producer's work and appreciate how little we really know.

Bravery Redefined: Overcoming the Fear of Flexibility

Acknowledging the role of the fixer is not only ethically correct for the decolonization of the profession, it also means better analysis of the potential benefits and pitfalls of the journalist-fixer relationship—a process that has begun in academic literature (Davies 2012; Murrell 2014, 2009; Palmer and Fontan 2007; Palmer 2108, 2019; Plaut and Klein 2019; Baloch and Andresen 2020). Murrell has highlighted that the fixer's role includes editorial input with obvious advantages including the suggestion of story ideas that might otherwise not be picked up (Murrell 2009: 11), but this scenario of supplying off the beaten track story ideas is more likely to be the case when the fixer is on a retainer and is starting to act more as a local producer. Murrell found that, to underscore the difference, BBC personnel who saw their local colleagues in this way would use the term "local producer," not "fixer" (Murrell 2009: 13).

The other way that the fixer influences editorial is through "forming the horizon of the journalist" either through a consistent pattern of analysis or through the fixer's contacts (Palmer and Fontan 2007: 15). This is not just a potential pitfall for the individual journalist; it can impact multiple journalists at a time. Hannerz quotes correspondent Anna Husarska as observing: "One young Bosnian Serb . . . 'probably had more influence on the way the world views the Republika Sprska than any other person'" (Hannerz 2004: 155). Similarly, Lederman writes how important changes came about in the types of stories reported through the setting up of the Palestinian translation and fixer service (1992: 76–78). The horizon forming can also impact news organizations in the long term, as fixers tend to be part of the "handover" provided to incoming correspondents by their predecessors (Erickson and Hamilton 2006: 41). Palmer and Fontan note that journalists who do recognize the risks of "horizon forming" try to work around the danger of a limited horizon through using multiple fixers over time. However, for parachuting journalists, even if they are returning to a place they've been to before, they are concerned with getting *a* story in their limited time, not *which* story. This means fixer diversity will be of less interest to them because they are more interested in reusing a fixer who they feel confident will deliver the goods. Often these journalists are going to "hot spots" where there is a strongly set up fixer "industry" ready to provide "canned material" (Erickson and Hamilton 2006: 41).

The use of multiple fixers over time is also difficult to enact in places like Goma, where there are not a huge number of experienced fixers. As an INGO interviewee said to me: "There's not actually that many fixers, so you end up using the same fixers, so you limit yourself to their networks . . . a lot of reporting is very personality based" ("Interview E" 2014). This personality-based sourcing of information could include fixers using the same contacts for multiple journalists where there is no direct overlap of markets. Dutch correspondent Luyendijk gives an example of this in Gaza, where an interviewee who gives him great quotations and who was recommended to him by his fixer, asks something to the effect of "Would you like me to say what I did for the French TV reporters?' (2009: 129). If journalism is a profession aimed at producing new knowledge through its reporting, then this recycling is closing off an opportunity for new knowledge production.

However, it is important to note in the example from Gaza that the interviewee gives great quotations. The ability of the fixer to find such a good interviewee shows that fixers understand the needs of journalists, and this understanding is what gets them work. This issue has been studied, from the point of view of fixers, in Iraq, with Palmer and Fontan (2007) finding that fixers felt they understood news values very well. This issue has also been studied from the perspective of the foreign correspondent. Murrell's research in Afghanistan found that correspondents enjoy working with, and want to employ, people with an understanding of their needs, including an understanding of "exactly the kinds of stories" commonly appreciated by Anglo-American audiences. Importantly, Murrell notes: "Hiring these kinds of 'globalized fixers' in Afghanistan does deliver stories that will be in tune with UK sensibilities, but these fixers do not necessarily represent how many of the people think (or vote) in Afghanistan" (Murrell 2013: 76).[2] Palmer and Fontan also highlight a danger in fixers' proficiency with common story frames and news values: fixers believe they know what western audiences want in terms of stories, so they exclude information in their translations, or their interpretations of the situation, which doesn't fit the previously defined frame. However, Palmer and Fontan found this was only recognized as a danger by a minority of correspondents (2007: 17).

Of the six fixers I interviewed in Goma (five male, one female), most also identified themselves as journalists within a spectrum ranging from those that identified foremost as a fixer to those who identified foremost as a journalist. The two most experienced identified themselves foremost as fixers and estimated that they had worked on upwards of 300 stories each ("Interview H" 2014; "Interview I" 2014). The most experienced fixer estimated that he had worked with 40 western news organizations and reeled off the biggest names in the business ("Interview H" 2014). It is worth noting the most experienced fixers had been in the job for almost 20 years, almost as long as the conflict in the east of the DRC has been raging. Like other researchers I detected a strong sense from fixers, especially experienced fixers, that they understood western news needs, and this allowed them to both secure their employment but also, sometimes, move stories on.

One fixer said that he would often suggest to his employer how they might move the story on with new angles—a suggestion he usually brought up by mentioning that a previous journalist has already done that story recently ("Interview H" 2014)—signaling he understood the professional importance of originality. He said this is not something he would have done at the beginning of his career but felt comfortable doing this now after working with "very professional and very talented journalists" who had taught him a lot. He also knew it was easier for him to make these suggestions, compared to other fixers, because his jobs were all through word-of-mouth recommendation. Another interviewee talked about the way he had worked with journalists to develop the "orientation" for stories ("Interview C" 2014). This fixer mentioned several investigative stories he had worked on using terms that implied co-development of the story idea with the correspondent. However, this fixer also described having to work with correspondents with "fixed" ideas and the frustration felt when some journalists chose to believe he was a "bad fixer" unable to deliver the goods when he told them the situation had changed.

The fact is, despite highly experienced local journalist/fixers having the professionalism that would allow them to work with correspondents to present new stories, or new takes on existing stories, parachuting correspondents brief their chosen fixer on their objectives for the reporting trip well before arriving—so where does the idea for the brief come from? Often from existing stories. Hannerz notes that "one way which foreign correspondents of the rotating variety often prepare themselves for a new assignment is to study the stories of their predecessors. This seems like a mechanism for preserving existing storylines" (2004: 217). One of Armoudian's interviewees describes the journalistic practice of drinking together as a method of generating story ideas—with many of those ideas recycled from what's been in the press already (2016: 30), and Luyendijk (2009: 23) shows how this reading up and recycling happens via the "cuttings library" and via interaction with the same sources that the previous correspondent spoke to—often facilitated through other foreign correspondents or through local producers. All of these practices can lead to what Lundstrom (2002) terms geographic bias—where certain storylines get attached to certain places, and it can lead to fixed ideas that hinder new knowledge production through journalism. The result is seen in Palmer's (2019) research, which shows fixers having to do the same old stories over and over again. Palmer also highlights that even though the workforce producing global news is multicultural, the reporting produced is Anglophone. Palmer's interviewees called the preconceptions leading to repeated storylines "false knowledge" and described how preconceived storylines hindered nuanced understanding and relied on sociopolitical stereotypes (2019: 55). One fixer told me that a documentary team was so attached to the idea for a story they had seen in another documentary that they refused to believe the fixer when he told them that the situation had changed in the year that had elapsed since that first documentary was made ("Interview J" 2014). This may be an extreme example demonstrating an ahistorical, apolitical understanding of sub-Saharan Africa, but each of the fixers I spoke to said that western journalists arrived with an idea and it was their job to make that idea come about.

I wish to make this point about inflexibility perfectly clear because it is easy to underestimate its importance. In all journalism, you begin your investigation with some sort of idea as to what the story will be. The crucial difference in this situation is that everything about the structure of the reporting experience, especially when "parachuting," militates against the normal flexibility with which initial story ideas are treated when reporting, say, the education round for a national paper. Arriving in a place like Goma, you have limited knowledge and limited time, and in that limited time you have limited contact with locals and limited, if any, local language skills, so you can't eavesdrop in the marketplace or "read the graffiti" (Palmer and Fontan 2007: 21). You are under extreme stress, your fixer is doing everything possible to fulfill the original brief, and you will probably have sold that original brief, that tried and tested story for the western market, to your editors back home who are making this trip possible. Bunce (2015) found that one of the main restrictions on the content produced by correspondents is the need to sell stories to a western audience (2015: 48). Basically, everything, including your own psychology, is working against the likelihood of changing the story dramatically, for the last thing you want is to introduce more volatility into an already uncertain situation. A Congolese stringer for a major overseas news organization (also a fixer) described this situation, saying that a western journalist changing the predetermined storyline required an act of "bravery," and, more often than not, foreign correspondents knew their storyline was flawed, and acknowledged as much in their conversations, but they didn't change it ("Interview D" 2014). Luyendijk also admits to this, discussing his continuing shame at how he "slipped up as a journalist" and ignored "inconvenient data" because that wasn't the story he'd "come to do" (2009: 128).

Lundstrom (2002) argues that news operations must tackle the elimination of geographic bias in the same way news organizations strive to eliminate racial, ethnic, and gender bias. Geographic bias is not only negative for audiences but also hinders the development of local journalistic talent because local journalists are at an even greater disadvantage than their western counterparts when it comes to arguing with their western-headquartered editors about what the story is. One local Congolese stringer ("Interview D" 2014) mentioned the storyline frames expected in the west as a stumbling block in developing his reportage. This is a problem for correspondents everywhere, with editors tending toward existing storylines (Lederman 1992: 13; Luyendijk 2009: 6; Palmer 2019: 55), but it is more of a problem for stringers: "The treatment of stringers and the home office's perception of them helps create a self-fulfilling prophecy. Lacking resources, trust, and financial security, stringers tend to go for 'safe' and cheap stories" (Lederman 1992: 93). In finding the bravery to be more flexible with storylines and eliminate geographic bias we can not rely on consciousness alone, practices are crucial. This could mean developing the new practice of asking fixers for story ideas they wouldn't normally suggest, it could mean editors taking risks on local stringers, and it definitely means taking into account the pros and cons of existing practices around international and local workers.

Postcolonial Hierarchies

When *The Guardian* journalist Daniel Howden went to cover the crisis in South Sudan in December 2013, *The Guardian*'s official twitter account wrote: "Daniel Howden, first western journalist into South Sudan, reports on brutal descent into civil war." A storm of protest followed with tweets like:

> @howden_africa Do western journalists covering an African story bring something new, something we Africans didn't get?
> @howden_africa it's just a misplaced sense of white superiority. South Sudan has many courageous journos.

Commentating on Al Jazeera, Nanjala Nyabola, a Harvard Law School–based political analyst, argued that the implications in the tweet would make her trust the western journalist less, as it demonstrates a belief in a "hierarchy of knowledge . . . based on race" (2014: online). She went on to argue western media would continue to get coverage of African issues wrong because of "their inability to confront this unspoken hierarchy of knowledge and the barriers it generates" (Nyabola 2014: online). This hierarchy of knowledge, which has survived into the postcolonial era, interplays with existing professional hierarchies both in newsrooms and in the field. For example Bunce (2015) examined international newsrooms in Nairobi and found that while local news workers were highly valued, "ultimately, however, the Western voices in newswires enforced through hierarchical chains of command, prevailed over and above dissenting opinions of local journalists" (Bunce 2015, 50).

This racial hierarchy of knowledge also impacts the fixer-correspondent relationship. All the fixers I spoke to in Goma saw their job as important because they helped the foreign correspondents to create important stories and capture international attention, yet none described the relationship as a partnership, as that word was too laden with equality, an equality they did not feel in their interactions with foreigners. This finding in my case study site is mirrored by the extensive Global Reporting Centre research study that anonymously surveyed 450 correspondents and fixers about their relationship with one another. The study found a divide between the correspondents and the fixers, with correspondents significantly undervaluing the role of the fixer when it comes to story direction. This quantitative result was followed with qualitative interviews to try to discern why there was such a large discrepancy in role perceptions:

> In the mind of the journalist, the *fixer* often ceases to be an individual and is seen simply as a tool for getting the *journalist's* story. In reality, by suggesting interview subjects or locations or providing local context, the fixer *is* framing the story for the journalist, though the journalist still maintains the illusion of individualism and power. The fixer recognizes and operates within uneven power dynamics, but is not without agency. Fixers are constantly engaged in framing the journalist's reality, and thus shaping the story, but they often do so in ways that are unseen by the journalist. (Plaut and Klein 2019: 1699–700)

In a telling point that speaks to the depth of this issue of nonrecognition, Nothias notes that even among foreign correspondents who displayed postcolonial reflexivity "the role of fixers was seldom brought up spontaneously by interviewees in these discussions" (2020: 267).

As noted earlier some global news organizations go against this devaluing and are searching for local non-Anglophone perspectives. The question is how far this will go in eliminating geographic bias. In conflict zones where fixers have converted into stringers, their considerable professionalism has nonetheless been built up on the job by catering to preconceived western ideas of storylines. For example, one highly experienced fixer, who also worked as a stringer, said to me: "Today there is a moto [motorcycle taxis—the main form of taxis in Goma] strike—that is not international news" ("Interview I"

2014). However, it could be international news. It could spark a story about civil unrest in the light of continuing state failure and questionable democratic reforms. I believe that this fixer did not see it as an international story he could sell as a stringer because the state of the DRC's democracy/bureaucracy is not one of the key storylines that has interested western media. It is unlikely that precariously employed locals in borderlands will break hegemonic news values unless specifically nurtured to believe that part of their job is to provide difference—that will first require acknowledgment that geographic bias and racial hierarchies of knowledge are a problem.

Holding on to Status, Outsourcing Danger

It is worth remembering that, for the most part, the change to local journalists doing the bulk of the eyewitness reporting is motivated by economics, not by a philosophy of valuing local knowledge. Again, there is a parallel with INGOs as traveling expat experts, especially white ones, come in and command the attention from authorities, and from other elements of their own organization, that locals cannot (Kothari 2006: 16). Yet locals hold together offices in the long term while their expat superiors come and go. Much like stringers, what Hamilton and Jenner call " 'Foreign' foreign correspondents" (2004: 317) and what the BBC call reporters on "local terms" (Murrell 2015: 152), these NGO professionals are offered employment on pay scales different from those of their expat colleagues. Palmer's (2019) important work rightly calls out the ethics of underpaying fixers. What I wish to further highlight is that this poor ethical practice has repercussions in terms of northern knowledge dominance. Jackson's ethnography of development workers describes this process of offering different pay scales in depth, viewing it as one of the structures of globalization that disempowers people's knowledge in developing countries because, as one of his interviewees puts it, their knowledge "comes cheap" and is therefore not valued (Jackson 2005: 114).

Driven by economics, this uncritical process of hierarchical knowledge then takes on its own momentum as new practices are developed in incredibly difficult, complex situations. One such potentially problematic practice has already occurred in Iraq. Quoting research from Orville Schell, Murrell describes a situation where much of the basic reporting is now done by Iraqis, while writing and analysis is done by westerners. Murrell goes on to say, "If this were to become the enduring model for correspondence, then the traditional foreign correspondent would morph into a 'foreign affairs correspondent' who would proffer analysis while the practical eyewitness reporting would be done by cheaper means" (2015: 154). Murrell calls this the "final step in the outsourcing . . . of international news gathering" (2015: 154). However, is this not also the outsourcing of danger? The type of ethics being practiced must be acknowledged and questioned. At the very least, it is not a practice designed to build a sense of solidarity; how will this privileging of the western life impact the analyses offered by the "foreign affairs correspondent"? It is worth noting again that there is a similar trend in international development, with 80 percent of "all victims of violence, fatal or not, among the staff of western aid organisations in war zones residents of the country in question," yet "there is barely any discussion of the ethics of this in the humanitarian world" (Polman 2010: 153). A similar deadly dynamic exists in Peacelands. In Mali, during peacekeeping operations in 2014, African soldiers were the ones bearing the front-line risks. This led to European military officers asking "whether the UN leadership was racist, as it exposed those soldiers to risks that would be unthinkable for Western or even Asian troops," but those questions were asked "in private" (Andersson 2020: 36).

Intimately entwined with ethics is respect: How respectfully will the raw "outsourced" information be used when it comes to telling the story? Sorius Samura became a freelance video journalist when his homeland of Sierra Leone was besieged by violence. Originally an actor, but now an award-winning documentary producer based in the UK, Samura's first interaction with the BBC was handing over to a western foreign correspondent, free of charge, footage he had shot. Samura's motivation was wanting the story to "get out and be told," but he was extremely upset with how the footage was used:

> Most of what I had been questioning about the coverage of Africa was really confirmed there and then: the West didn't care about context. They told the story their own way. They don't show us respect; they don't trust us with our own stories; they just portray stereotypes. It's really frustrating. Sometimes you think, "Why the hell would I want to continue taking risks when someone else goes into that cutting room and does what they want to do with your material?" (Samura in Leith 2004: 316)

Sambrook is another academic who predicts the morphing of the foreign correspondent into the foreign affairs correspondent. He describes the foreign affairs correspondent's work as made up of "verification, assessment and analysis" of information supplied by others, including social media and the blogosphere. Sambrook argues this sort of computer-based reporting, driven by the limited production deadlines of the internet age, offers no time for a "prolonged process of firsthand engagement" where the correspondent can "'unlearn' . . . preconceptions" (2010: 36–38). One former Crimea correspondent, now practicing foreign affairs correspondence, describes it as "the practice which used to be known as plagiarism," and comments that there is a higher degree of reliance on specific viewpoints, including from national media, which may not be as balanced as an outside observer (Armoudian 2016: 5). Williams (2019: 184) describes coverage of Syria as "desk-bound," reliant on "videos, NGOs, press releases and blogs" as well as Skype interviews. He points to several problems, including insufficient verification, that have led to false reports and, disturbingly, the fact that the reliance on social media also amplifies the information war. "As a weapon of war, social media has been used to generate images of death, destruction, killings and atrocities to influence opinion and intimidate opponents" (Williams 2019: 184).

The impact of foreign affairs reporting raises important new concerns that must be dealt with as part of decolonizing the profession. In a seminal anthropological work on foreign correspondence from the early 1990s Pedelty (1995: 203–17) doesn't use the term "fixers," but the structure of the relationship is easy to identify, with job insecurity, greater danger of death, and underappreciation of their skills described by local El Salvadoran journalists employed by international reporters. After 30 years Williams (2019: 184–85) argues that the increased reliance on fixers is one of the features of war reporting in the global era and that it is no longer an unrecognized and thankless task. This may be true, yet is this recognition enough? Remaining in a position subordinate to the "Muzungu" must be hindering the Congolese fixer who is also a freelance journalist. Two of my interviewees complained that Congolese were not given the trust and opportunity by international news agencies to file stories off their own bat and build up their own career ("Interview D" 2014; "Interview I" 2014). The clearer incorporation of fixers and stringers into the international news environment may actually end up, at least in the short term, consolidating the current "star" positions of firefighting correspondents from Anglo-American news organizations. This is because "a very effective way of consolidating a powerful position in a field of relations is to institutionalise it" (Nash 2016: 173). Harris and Williams (2018: 60) note that "local knowledge is required for parachute journalism to function effectively, and stringers and fixers contribute the servicing of Western star reporters."

The fact that the foreign correspondent maintains a celebrity position (Sambrook 2010: 7) is shown by the resources used in parachuting, which, while less costly than maintaining foreign bureaus, are still huge when viewed on a per issue/crisis basis and when compared to what is spent on stringers per story.[3] The focus on stardom is further seen in the jockeying undertaken by "famous names within a media organization [who] compete to cover the headline stories—whether or not those journalists know anything about the subject" (Lederman 1992: 138). Audiences are then socialized to understand that a disaster is "major" when the stars are there (Moeller 2006: 187) and these practices then become self-perpetuating. Furthermore, to borrow Keane's (2004) terminology, this is color-coded stardom; while modern correspondents may be relationally, rather than physically, white, it still entrenches a framework of western/northern superiority.

At the end of the day, while much in the field has changed, the continued use, *and celebrity-like promotion*, of parachuting western foreign correspondents reinforces the hero trope, which is further used as a story frame to fill the gap in content for journalists bereft of local knowledge. This was noted in Middle East reporting in the early 2000s: "There was endless reporting of, 'this just happened. A rocket came over, we barely got missed.' There was too much I, I, I'" (Pintak in Ricchiardi 2006: 47). More recently Palmer (2018) has looked at Anglo-American coverage post–September 11 to reveal how the danger faced by correspondents has become a key storyline of these new-style wars. Drawing on the tragic death of correspondent Marie Colvin in Syria, Palmer writes that her death was used by networks to "bolster their rather simplistic representations of the Syrian opposition" (2018: 156). Thus, "Colvin's final contribution to 'the story' of Syria continued after her individual end, signalling the slippage between the conflict correspondent as storyteller and as cultural sign, made to speak beyond her own intentions" (Palmer 2018: 132).

Another Case for Affirmative Action?

This chapter started with acknowledgment of the danger journalists face because many of the changes to journalism in conflict zones have been driven by changing risk dynamics. These changes can't and shouldn't be ignored. At the same time it must be recognized that turning a local crisis into an issue about the security of the correspondent is another form of erasure journalism. Even now, at the time of writing this chapter, there is an example in the Australian media with the story of the arrest of Australian journalists in Malaysia, overshadowing the story they went there to do—a story of "intrigue, corruption and multiple murders, stretching from the streets of Malaysia's capital Kuala Lumpur, to Switzerland, France and the US as well as Hong Kong and Singapore, all the way to Australia's doorstep" (ABC 2016). Of course the arrest of the correspondents is important, but power relations cannot be ignored; focusing on the arrest makes the impact on the western life the central and unusual drama and normalizes corruption and arbitrary arrests for other lives.

In chapter 2 it became clear that affirmative action in terms of word choices (the gorillas, guerrillas fiasco) may be appropriate. Another case for "affirmative action" can be made here. The investigative journalism team from the ABC Four Corners program was obviously tackling a type of world connections story that international correspondents should be pursuing. The story of their arrest fits the overall narrative because part of the narrative is the climate of fear and political repression in Malaysia. But the correspondents' arrest was not needed; there were many other local examples to use.

I realize this seems like an extreme omission and goes against every storytelling instinct, but complex stories are difficult for audiences to engage with, so when there is an opportunity to latch on instead to a recognizable element of the story that is more easily accessible from their primary frameworks, it is perhaps natural that the audience's attention will focus there. By including the hero/adventure trope we could be encouraging our audiences to focus on the comfortable reaffirmation of the white man's burden instead of looking at the complex story the correspondents (and their local partners) are trying to tell. Innate human interest in drama is also a consideration here. I'm sorry to say the story of our arrest in Sudan was among the most-read articles on www.itbeganinAfrica.com. This isn't surprising. It is very human to be attracted to personal drama, and our audience had a preexisting relationship with us from the previous six weeks of our trip. Steve and I tried to separate our arrest story from the more important story of the plight of the people in Kosti by posting three different stories from that day and the arrest story as a blog—not in the main journalism part of the website. The post in the blog is also only about the arrest in the beginning; the main thrust of the story is the intrusion of the security apparatus into ordinary Sudanese life and the contrast between the ethos of the state and the ethos of the people we met. Is the story of the arrest saved from being merely a yarn of interest, and does it become a reflexive piece of journalism thanks to this broader discussion of complexity of Sudanese culture and society? I can't answer that. Nor can I know for sure whether or not the white hero / white man's burden trope was activated. All I can say for sure is that it is possible. The trope may indeed have been activated by the dangerous journey frame and that people paid more attention to our tale of danger than to the stories of the Sudanese whom we were there to report on.

Consider again "The Congo Connection." As mentioned in chapter 1, this difficult story of connections is confused through a number of different framings, one of which is the "foreign correspondent's difficult journey." Unfortunately, that frame is given even more emphasis and linked more closely to the hero trope by the presenter in the studio, who bookends the report with the words "Eric Campbell, reporting from one of the world's more difficult destinations" ("The Congo Connection" 2009). Given that an anchor's few words can have a "disproportionate effect on perceptions" (Lederman 1992: 300) and that the hydraulic effect of frames involves one interpretation driving out other possible interpretations, are these words really helpful or necessary in an extremely difficult piece? Furthermore, those words suggest that the danger to the *foreign correspondent's* life is the central and unusual drama of the story. This subconsciously references the biopolitical divide, for it creates a distinction between those whose lives are *expected* to be in constant crisis (the poor locals), and those whose lives are not. Palmer points to the same dynamic taking place in the Middle East reporting post–September 11, where the capture and death of conflict correspondents are used to create melodramatic narratives that at the same time "frame some deaths as more politically and ethically important than others" (Palmer 2018: 156).

A final pitfall we can discuss here is the way the hero/adventure frame flattens the distinction between the importance of "getting there" to tell the story, and the more important achievement of "understanding there," to tell a story that strengthens the global public sphere. The distinction between those two concepts can be illustrated by considering the work of Wilfred (pen name Peter) Burchett, an Australian foreign correspondent, the first journalist to report on the aftereffects of Hiroshima, and the first to understand the "*experimental* nature of this first use of a nuclear weapon against people" (Pilger 1986: xii). As noted in a eulogy for Burchett, "It was a considerable ordeal to reach Hiroshima . . . but it was an infinitely greater accomplishment, back then, to *understand* the importance of Hiroshima" (Allman in Pilger 1986: xii). Burchett's report doesn't mention his difficulties getting to Hiroshima at all. He does use plenty of personal descriptions, but only in an attempt to increase understanding, as in paragraph 3 when he writes, "In this first testing ground of the atomic bomb I have seen the most terrible and frightening desolation in four years of war. It makes a blitzed Pacific island seem like an Eden" (quoted in Tanter 1986: 18). Yet Burchett's editors at the *Daily Express* couldn't resist the hero frame. Their introductory paragraph for the story reads:

> Express Staff Reporter Peter Burchett was the first Allied Reporter to enter the atom-bomb city. He travelled 400 miles from Tokyo alone and unarmed, carrying rations for several meals—food is almost unobtainable in Japan— a black umbrella, and a typewriter. Here is his story from—HIROSHIMA. (Quoted in Tanter 1986: 18)

No doubt the editors thought this dramatic, black-umbrella-carrying (talk about playing to a stereotype!) hero protagonist would draw people in to read the report, and perhaps there was no damage done to the "warning to the world" (Tanter 1986: 18) Burchett provided. But this is the case with one exceptional story by an exceptional reporter, and the possibility for damage was still there. Ultimately, that introduction, even if it didn't damage the message of *that* story, did add yet another strand to the weave of the foreign correspondent hero trope. It is that strong weave, built up through repeated uses of the hero frame in individual pieces of journalism, that has helped to fashion today's trends of celebrity firefighter correspondents, hidden fixers, and the devaluing of local knowledge, which we must value, if we are to understand the specificity and complexity of borderlands.

5

Reporting the Local and Global
of Borderlands

New wars are fought by networks of state and nonstate actors that
are both global and local. Indeed, global circuits are critical in both
political and economic terms. Often the individual combatants are
recruited from rural areas or from recently arrived rural-urban
migrants. However, they are incorporated into transnational
networks, whose political communications and financial infrastructure are necessarily city based.

—Kaldor and Sassen 2020: 7

Early on in this book I asked this question: How is it possible that Australian
voters *couldn't* understand the will to live or die attendant in the boat
journey to Australia? Part of the answer given was the degradation, in our
eyes, of people's humanity—the acceptance of bare life as enough. Yet tragically, it is that degradation of humanity in regions subject to ongoing conflict
that creates the will to live—or die trying. Mbembe has aptly termed these
regions necropolises and described how the permanent condition of "being
in pain" (2003: 39) leads to a search for agency even if that agency is death.
He writes of

fortified structures, military posts, and roadblocks everywhere; buildings
that bring back painful memories of humiliation, interrogations, and
beatings; curfews that imprison hundreds of thousands in their cramped
homes every night from dusk to daybreak; soldiers patrolling the unlit
streets, frightened by their own shadows; children blinded by rubber
bullets; parents shamed and beaten in front of their families; soldiers uri-
nating on fences, shooting at the rooftop water tanks just for fun . . . border
guards kicking over a vegetable stand or closing borders at whim; bones

Borderland. Chrisanthi Giotis, Oxford University Press. © Oxford University Press 2022.
DOI: 10.1093/oso/9780197565797.003.0006

broken; shootings and fatalities—a certain kind of madness. . . . As Gilroy notes, this preference for death over continued servitude is a commentary on the nature of freedom itself (or the lack thereof) . . . death, in this case, can be represented as agency. For death is precisely that from and over which I have power. (Mbembe 2003: 39)

The condition of living in these necropolises is not understood by the majority of the Australian public and yet they are not other worlds. Mbembe describes their existence as part of the "new geography of resource extraction" (2003: 34)—they are intrinsically part of the global economic sphere that connects us all:

> The extraction and looting of natural resources by war machines goes hand in hand with brutal attempts to immobilize and spatially fix whole categories of people or, paradoxically, to unleash them, to force them to scatter over broad areas no longer contained by the boundaries of a territorial state. As a political category, populations are then disaggregated into rebels, child soldiers, victims or refugees, the "survivors," after a horrific exodus, are confined in camps and zones of exception. (Mbembe 2003: 34–35)

In short, necropolises are modern testaments to the observation of Walter Benjamin that "there is no document of civilization which is not at the same time a document of barbarism" (Benjamin 1968: 248). Brutal methods were used in the extraction of resources that built the metropolises of empires and brutality accompanies resource extraction today. The "civilized" act of placing a diamond on a loved one's finger has its corollary in devastation in Sierra Leone and Liberia. The act of using precious-mineral-dependent smartphones has its corollary in war in the DRC. But this is only one part of the analysis. These zones are also productive of a more subtle and complex economy based on the "humanitarian-security nexus" (Andersson 2018: 424) not only visible in the borderlands of war and conflict but also repeated in the securitized border zones at the edges of Europe, North America, and Australia's "pacific solution" to warehousing refugees. Kaldor and Sassen remind us that

> although casualties (deaths) are, on the whole, lower than in the industrial wars of the twentieth century, the typical characteristic of contemporary wars is expulsion. Political control is established through the forced

eviction of those with a different identity or those who disagree. Every year, the grand total of refugees multiplies. (2020: 9)

This humanitarian-security nexus is one of the major journalistic stories of our time. Yet, as noted by Powers (2018: 142), even simple stories that involve the "dynamics of international humanitarian provision" struggle to get onto the news agenda, as they are seen, by western head-office editors, as "outside scope"—too complex and too slow moving to be worth the investment of foreign correspondence. Nothing could be further from the truth.

In telling this story it is important to understand the global currents that shape borderlands. However, the way these forces manifest locally to create *glocal* realities is also crucial. The glocal nature of social vulnerabilities has already been raised as a key issue in sociology by Beck (2009: 163). Likewise, Agier (2016: 125) states that the local/global dynamic is at the heart of the cosmopolitan condition that should be understood by anthropologists in borderlands. This chapter argues foreign correspondence too must take up this challenge. To help in this task this chapter first considers some key arguments from theorists who look at the global dynamics of borderlands and then introduces three case studies sparked by the key concepts from previous chapters. Analysis around the issues of artisanal mining, "NGO-speak," and the way fixers get caught up in the "us and them" discourse will be used to discuss how reporting practices can hinder glocally focused reporting, and what can be done to move past these blocks.

The Global Border

The border as an analytic tool helps focus the mind on processes of globalization that are also the processes of new wars. This is visible in recent literature from international relations and migration where Mbembe's theorization around necropolitics has been used to expose the warehousing governance around "the racialised constitution of the 'living dead'" (Aradau and Tazzioli 2019: 204). This governmentality around "bare life" is especially visible in border zones; however, authors Aradau and Tazzioli (2019) go further. By contrasting two different European border zones (those of Greece and Calais) they argue for new analytic categories that consider how the governance of refugees tends to incorporate them into extractive industries (e.g., through data harvesting) or else subtract them from the public sphere.

Journalists too have used border processes to illustrate the dynamics of new wars. Maria Armoudian interviewed a journalist (who remained anonymous) on her choice of stories around the Syrian conflict. The journalist, given the label PC10, spoke of a story where she described how people would risk being shot by snipers as they crossed from government-held territory to rebel-held territory but would make this treacherous journey every day in the quest for cheaper groceries. Returning to the border crossing, they would run as fast as they could to avoid the snipers yet not so fast that they would drop the groceries. For PC10 this tragic vignette represented the adaption to the new reality of the drawn-out war. In another story she focused on the rebels' use of tunnels under government territory used to blow up buildings, this representing that they were willing to destroy the country to continue the fight. For PC10 even as early as 2012 it was clear that the conflict was resembling Iraq—another unending war (2016: 17–19).

My case study is the eastern border of Democratic Republic of Congo. Specificity matters. However, there are also global dynamics that need to be understood. In *Cities at War*, a collection of essays from postcolonial and post-Soviet countries, key trends are seen. In a borderland town like Juarez, Mexico, transnational corporations are protected by government, and the role of the private security industry becomes a key concern (Martin 2020). This is similar but different to the importance of transborder trade to the city of Goma, controlled by military-economic elites (Büscher 2020). The social geography of Karachi, and in particular the increased enclave-making and securitization, is impacted by the borderland on the other side of the country (Kaker 2020). This is again similar but different to the way Bamako's geography was changed when it became headquarters for remote management of peacekeeping and development operations in the north of Mali. In Bamako, the particular form of change is modeled on Kabul, with Mali called Africa's Afghanistan (Andersson 2020). This hub-and-spoke system of peacekeeping, also known as archipelago intervention, is similar to the "islands of stability" framework that I was introduced to in Goma where the goal, rather than peace, was to secure certain high-value regions.

The "islands of stability" approach in the east of the DRC was criticized as concurrently creating "swamps of insecurity" (Vogel 2014; Cooper 2014). Moreover, Vogel (2014: 5) points to the political problem created by this strategy in that "insecurity and imbalances in security provision create tensions and security voids on local levels." In September 2021 the United States withdrew troops from Afghanistan. In the lead-up to that withdrawal western powers and media were shocked at the collapse of the Ghani government and the rapid "fall of Kabul" to Taliban forces. Kabul had been the high-value, protected city of the US military security, and US State Department nation-building, strategy. In the hub-and-spoke system of security, the separation of parts of the country from each other was mirrored in the enclavization within Kabul, powerfully symbolized by ubiquitous blast walls used to protect high-value buildings from suicide bombers. As journalists, we must ask questions about the overall wisdom of these conflict strategies that are powerfully supported and maintained by global discourses.

Kaldor and Sassen frame different cities as existing in different global security cultures, drawing particular attention to the war on terror and the liberal peace.

The war on terror is about the use of military force to attack nonstate actors—it is the war of the manhunt, *par excellence* asymmetric war. The war on terror involves a sinister twenty-first-century combination of intelligence agencies, private security contractors, and regular military forces . . . warfighting results in very high casualties, especially among civilians . . . the prevalence of these external interventions is characterized by partitions, security zones, and extensive surveillance . . . a process of *enclavization. . . .*

The liberal peace is about stabilization after a formal peace agreement and involves a combination of peacekeepers, civilian aid agencies, and NGOs. The liberal peace model of security is about implementing a peace agreement in which the participants in the peace agreement are the actors of new wars . . . it cannot control the ability of armed actors to prey on ordinary people, because the armed actors are the key agents in implementing the agreement . . . this model of security can be seen as a way of living with perpetual war. The walls and checkpoints become emblematic of threats, and physical distance increases the suspicion and fear that is the bedrock of identity politics. This model of security is also expensive . . . [it] creates an industry of insecurity that may self-perpetuate and add to existing insecurities. (Kaldor and Sassen 2020: 10–13)

These security cultures are not mutually exclusive. Kaldor and Sassen highlight cities like Kabul, Baghdad, and Bamako where the liberal peace coexists with war on terror and where insecurity is pervasive (2020: 13). What is more worrisome is the fact that these security cultures may be co-constituting. Goma is given as an example of a city where the liberal peace is not overshadowed by the war on terror and where urban capabilities help to create the city as a safe haven in the midst of war (2020: 14). I question how long this can last. As noted earlier, the security culture of the liberal peace is attendant with physical distance that increases suspicion; this can be taken advantage of by violent actors, as happened in Mali (Andersson 2020).

In 2020, in the east of the DRC, events took place that on the surface seem contradictory but in reality could be perfectly aligned:

- The five-star Serena hotel opened in Goma.
- Civilian casualties at hands of militants doubled.

Fighters from the Allied Democratic Forces (ADF), a group originally based in Uganda that is increasingly connected to global Islamic extremists, were the major difference in the increase in casualties and responsible for a third of the deaths. They are operating largely in the area of North Kivu (the state of which Goma is the capital) and Ituri province (the state immediately north). While five-star luxury is not immediately associated with war zones, the Serena hotel's website highlights the potential for terrorist-style attacks with the home page welcome video prominently including vision of the undercarriage of cars being swept with mirrors before entering the hotel (Serena Hotels 2021). In April 2021 multiday protests against the UN mission erupted in Goma, Beni, and Butembo (Al Jazeera 2021). These initially largely peaceful protests were the result of anger and frustration at the increased actions of the ADF and MONUSCO's seeming nonresponse. Writing these words from a cafe near my university office, I make no pretense of being able to understand the unfolding situation. I just note there are obvious, changing dynamics that are both local and global.

Doing justice to the local, by taking into account the global, is necessary to represent the dynamics of new wars and to search for ways out. Kaldor and Sassen (2020) make the point that not enough attention has been paid to the new identities and capabilities that are created at the local urban level in the midst of cities at war. To play its part, foreign correspondence will require heightened sensitivity to the local-global dynamic—a sensitivity that is in itself a process of decolonization, for it opens the space to understand the political present of the borderland. Agier argues:

The local/global relationship is the "basis and precondition" for the cultural dynamic observed at a given place. It creates an in-between situation, embodied in places and moments of encounter, exchange of ideas, misunderstandings and new political conflicts, giving rise to new symbolic corpuses and new political subjects, increasingly informed of—and formed by—their cosmopolitan condition. (2016: 125)

In considering how this glocal sensibility can be incorporated into foreign correspondence, this chapter's case studies focus on different elements of professional practice. The first case study raises issues around story-frame selection and avoiding the lacuna effect, the second analyzes glocal dynamics impacting journalist-source relations, and the third considers how best to communicate the glocal nature of the growing dangers of conflict reporting. Each example illustrates the need for decolonization work before international reporting can reach its potential and explain the glocal realities of borderlands to a global public sphere.

Artisanal Mining and the Lacuna Effect

In 2014 the timing of my visit to the DRC coincided with a significant spike in western media and policy attention to conflict minerals. Not long before, important provisions in the US Dodd-Frank Wall Street Reform and Consumer Protection Act had come into effect. Section 1502 of the Dodd-Frank Act requires companies selling goods in the United States to guarantee their products are free of conflict minerals. Media reports told me that the major issue in the DRC was conflict minerals, and, what's more, the solution to this issue will have to be achieved through western intervention and action. Known colloquially as "Obama's law" by artisanal and small-scale miners (ASMs) in the DRC (Vogel, Musamba, and Radley 2018: 79), this decision on the other side of the world had major implications for their lives because the concerns of the ASM community had not been listened to ("Conflicted" 2015).

The insertion of Section 1502 into the Dodd-Frank Act did not happen overnight. In fact it could be argued that its genesis came in 2000 with the story that brought global attention to conflict minerals. It was published in the *New York Times* and coordinated by former *Washington Post* Africa correspondent Blaine Harden, with input from four other foreign correspondents, reporting from Sierra Leone, Congo, Angola, and Botswana—it also visits the diamond bourse in Antwerp, seen by many as a diamond smuggler's dream, and reaches to where the story, for most of its readers, is no longer foreign news: an electrician from Staten Island stands with his fiancée at a diamond counter on Forty-Seventh Street in Manhattan, a few blocks from the New York Times headquarters, eyeing a $5,500 ring. (Hannerz 2004: 185)

Hannerz documents this story as an excellent example of foreign correspondence taking up the challenge of reporting complex interconnectedness. And I agree. So when I first heard the words uttered by an INGO worker in a scathing tone, "If I read one more story about conflict minerals," I was shocked. And to be honest, I was dismissive. I thought it was a case of sour grapes, of this worker wanting more stories on the INGO's particular issues, but in Goma I was privy to four discussions on the media representation of conflict minerals. Their purport centered on the fact that the dominance of this one element simplifies a complex situation by ignoring the other aspects driving the conflict (discussions August 30 and September 3, 2014). In addition, its salability as a story in the west means journalists are not interested in reporting on other issues of importance for local Congolese ("Interview E" 2014; discussion September 4, 2014).

In short, the key issue was a lack of "ground truth," that is, detail concerning people's lived reality in the DRC. In particular, the focus in the foreign media on western intervention and sanctions meant scant attention paid to the issues surrounding ASM not involved in the conflict economy (Fahey 2009). Researchers now argue that "the incompatibility of ASM with corporate-regulatory structures prevents a purely positive impact of traceability and certification" (Vogel, Musamba, and Radley 2018: 74) and moreover give examples where the implementation of ethical tracing in the tantalum, tin, and tungsten trade has driven "miners into either unemployment, gold mining, other precarious occupations, or—given many miners are demobilised combatants—back into armed group recruitment in certain cases" (Vogel, Musamba, and Radley 2018: 79). If more attention had been paid by journalists to the actually existing structures around the mining industry, questions should have been asked as to how the international tracing systems would be integrated into the existing ASM sector. Why did this obvious line of investigation get largely overlooked? Potential answers exist in considering the lessons learned from the discussion around the Africanist discourse in chapter 1.

If we take the article by Harden et al. (2000) as an early example of the conflict mineral story, we can identify an overall frame, especially strong in terms of selection and repetition, that *wars in Africa are driven and prolonged by illegal mineral trade*. However, this is an unusually long piece (described as a special report and given front-page billing), and there is room in its more than 5,000 words for a strong secondary frame to emerge of *western culpability*: arms sales, corporations' failure to take responsibility for where their products come from, and consumers' lack of care for where our diamonds come from. A small, third storyline also emerges, *African governance and responsibility*, with Botswana given as an example where diamonds have not been a curse. Unfortunately, this third potential frame is underdeveloped compared to the previous two, not least because there are no quotations from Botswanan government officials. Instead, as so often happens, a western analyst is given voice on this topic.

Once the *New York Times* put conflict minerals on the media agenda through a major story, it is natural that other media organizations would start reporting on the issue (Golan 2006). However, the alternative secondary framings identified in the initial exposé have not been brought to the forefront by journalists following up the story. Instead, sedimentation into the cause and solution frame of "mineral riches fuel wars and western culpability needs to be addressed" has occurred.[1] The dominance of this particular frame has been helped along by the interests of INGOs and in particular a well-connected INGO called the Enough Project. This organization campaigned strongly for the inclusion of Section 1502 in the Dodd-Frank Act (Vogel, Musamba, and Radley 2018; Laudati and Mertens 2019) and has also been credited with catalyzing the connection between "Congo's mineral economy, Western consumers and the brutal and widespread use of sexual violence in the country" (Laudati and Mertens 2019: 64). However, a messy fact that doesn't fit into the dominant narrative is that there are hundreds of thousands of nonviolent artisanal miners who rely on the current arrangements for their survival. The World Bank estimates that 10 million people—12.5 percent of the 80 million population of the DRC—derive their livelihoods from ASM (Bakumanya and Tounsi 2020).

Here surely is the lacuna effect in action. As discussed in chapter 1, there is "erasure journalism" at work that removes Africans from their stories and replaces them with westerners; the artisanal miners are practically nowhere to be seen. As importantly, the lacuna effect naturalizes an appalling lack of curiosity and questioning; the vague representations of how the future mining system would work with the current mining system were not scrutinized by journalists for their lack of detail. Furthermore, as predicted by Easterly (2006) in *The White Man's Burden,* when one is focused toward the west, grand plans for positive change can leave devastation in their wake. Vogel, Musamba, and Radley write:

> The fixation on consumer protection above local dynamics results from policy that was designed as a response to western-led advocacy efforts. It gravitates around the question of how Congolese artisanal mineral production—which is embedded within global production networks—can be fed into a verifiable supply chain in order to provide western consumers with the leverage to buy products not associated with conflict activity (and exclude the opposite). (2018: 77)

Late in its campaigning for Section 1502, the Enough Project acknowledged the potential harms to legitimate artisanal mining (Bafilemba, Mueller, and Lezhnev 2014: 10–11) however, the organization argued, with the help of a particular case study, that the western mine taking over would provide alternative development support for the affected local community that, at its peak, was 10,000–15,000 strong, with a base of 2,500 artisanal miners. The reality since then has turned out to be much more complicated. The Toronto Venture Exchange–listed, Mauritius-based, tin-mining multinational Alphamin Resources got off to a rocky start. In 2016, four years after obtaining a mining license for the area and stopping the artisanal mining, much was promised for the future but, by its own account, the company had only provided road and telecommunications infrastructure. A small area that was meant to be set aside so that artisanal miners could continue work in an accredited conflict-free zone was still to be set up (Alphamin Resources 2016). Indeed it is now clear that by 2015 Alphamin Resources had decided it would no longer support artisanal mining on the site, and in 2017 a plan was made to buy out and relocate the existing miners and mine owners (Fahey and Mutumayi 2019). In a report, in part produced in response to continuing criticism of the mine's operations, the researchers note that the process of relocating around 900 people happened remarkably peacefully and that 50 artisanal miners chose to stay on and work for the company. They also say that the company has, overall, created employment for 1,500 Congolese people during the final phase of construction (Fahey and Mutumayi 2019: 1). Yet there is no explanation of how the ongoing mine operations will compare to the "thousands of people" directly and indirectly employed in the preindustrialized production processes (Fahey and Mutumayi 2019: 8). It is also clear that instead of undertaking the complex task of integrating local artisanal mining into global production processes in a win-win situation for all, the mine operators have focused their efforts on a more charitable frame of reference. They have set up a grant body that is working with local communities on projects like schools, small businesses, and soccer pitches (Fahey and Mutumayi 2019: 25). Vogel, Musamba, and Radley (2018) found in their interviews that there is a high degree of skepticism about any claims that conflict-free industrial mining comes with regional development opportunities.

The conflict mineral narrative also reveals further issues for decolonization work because of the role of the sexual savage discourse. International relations scholars Laudati and Mertens (2019: 73) describe how "the focus on rape in mining areas obscures other issues of importance" for locals like the environmental degradation caused by industrial mining. Furthermore, constructing ASM "as *the* space of violence linked to minerals and sexual violence" ignores the evidence of sexual violence at state and corporate mining sites (Laudati and Mertens 2019: 73). The Alphamin example also highlights the continued strength of the colonial era discourse that holds that western exploitation of natural resources is beneficial to African development, despite the ample examples from history to show the opposite is the case. This discourse reared its head again with the fall of Mobutu, when "the familiar vision of the Congo as an untapped source of wealth gained renewed currency" (Dunn 2003: 167) and allowed "western business interests to continue to exploit the Congo's promised 'potential' in the name of 'development'" (Dunn 2003: 170). Of interest is the fact that the artisanal miners' cooperative joined with a Mai-Mai rebel group to attack the Alphamin mine site in July 2014 (Pole Institute 2014). The attack, and in particular the partnership between locals and the rebels, reveals how *material* life can show up incongruities with *representations* of space—after all, according to media representation of the eastern DRC, the rebels are not meant to be partners with ordinary locals, yet "in many cases, effective resource extraction involves cooperation with local authorities and civilians" (Laudati and Mertens 2019: 70). Moreover, Sassen points out that international mining processes hugely complicate the issue of governance. She states that when sovereign national territory is turned into land for usufruct, "this process, at least indirectly, degrades the governments that sold and leased the land. The eviction of farmers and

craftspeople, villages, rural manufacturing districts, and districts of agricul-
tural smallholders similarly degrades the meaning of citizenship for local
people" (2014: 82–83). This outsourcing is in direct contrast to the strategy
deployed under Mobuto's kleptocratic regime, where gold was bought from
artisanal miners through the Zaire central bank—a strategy that some in the
DRC are asking the government to revive (Bakumanya and Tounsi 2020).

Agier builds on Sassen's concept of the denationalization of the economic
space to argue that humanitarian intervention is also a delocalization of the
political space (Agier 2016: 52). The liberal peace, which is meant to help secure these mining operations, adds another layer of complexity to notions
of sovereignty as security of, and between, mining sites and communities
has been outsourced. The "islands of stability" strategy can be seen as one
of risk mitigation, and yet the dynamics of new wars push peacekeeping
missions into ever more drawn-out and expanding operations that create a
new governance structure. Peacekeeping missions are meant to support local
governments, yet as they extend indefinitely the internal contradiction of
foreign occupation built into the practice becomes ever more evident. Yee-
Kuang Heng notes:

> UN transitional authorities have replaced the previous "trusteeship"
> colonial-style system which has become politically unacceptable. Going
> beyond traditional peacekeeping, this system reconstructs governance and
> security and some state powers are temporarily assumed. Yet it is hampered
> by lack of resources and the contradiction between *ends* (legitimate govern-
> ance) and *means* (foreign occupation). (2006: 82)

Foreign occupation becomes a key global dynamic of the borderland's local
politics and helps legitimize the global discourse that defines the problem
(or risk) as being conflict mineral *sites*, rather than conflict mineral *processes*.
Appadurai argues: "The locality (both in the sense of the local factory or site
of production and in the extended sense of the nation-state) becomes a fetish
which disguises the globally dispersed forces that actually drive the production process" (1990: 306–7). This emphasis on locality obscuring transnational
processes is seen in "The Congo Connection" (2009). In this episode foreign
correspondent Eric Campbell works toward a type of global connections story.
To his credit, while talking about the devastation enabled by warring groups
exploiting minerals, Campbell does not ignore the importance of the artisanal
mining for the survival of hundreds of thousands. Unfortunately, voice on this

issue is given to a western analyst. Like the Harden et al. story, Campbell also makes the connection to western consumers. Furthermore, he makes a specific connection to Australia, saying, "Australia is the only other country to have large reserves of coltan, known in Australia as Tantalite, but the Australian mine has stopped working, claiming it can't compete with cheap African exports" ("The Congo Connection" 2009: @17m45s). This could be the beginning of an interesting discussion about global production processes, but it stops at a discussion of production localities—just as Appadurai's production fetishism predicts. The Australian mining industry's connection to the DRC does not end with competition over the location of production. Stopping the story at this point obscures the fact that at least one major Australian mining company holds mining concessions in the DRC and prefers to hire seasoned Zambian workers rather than locals (Rawlence 2012: 204). The production fetishism also hides the people involved, the globalizers who make the processes possible, like the young Australian mining executive whom I observed in 2014 having a Skype meeting on his iPad thanks to the excellent Wi-Fi at Goma's exclusive lakeside Ihusi Hotel.

Where have we come to at the end of this example? I wish to be clear that I'm not saying Section 1502 should not have been passed. This isn't an argument about the rights and wrongs of policy, or even about the when and what of western intervention. As noted in the introduction, "A global public discourse does *not* arise out of a consensus on decisions, but rather out of *disagreement* over the *consequences* of decisions" (Beck 2009: 59). The role of journalism is not to provide *the* solution to any specific problem but rather to create knowledge through discussion, to ask questions, to flag potential problems, and sometimes to raise potential answers. The point I wish to emphasize now is that I use the terms "frames" and "discourses" deliberately. The potential destruction of the artisanal mining economy was mentioned in "The Congo Connection," but it was buried in the primary animal frame of "saving the gorillas" and the secondary frame of "sexual violence and chaos." Furthermore, although connections were made to Congolese governance, again, as with the *New York Times* piece, no local government spokespeople and no civilian opposition groups were interviewed on the issue, only a western analyst. This weakens the argument because a discussion of local governance is framed by a white expert, which activates the white man's burden trope, and also privileges global over local solutions. The issues of exploitation (minerals, land, people) in borderlands are not local or global issues; they are *glocal* and need to be framed as such.

Journalist-Source Relationship or Muzungu-Victim?

In chapter 3, I made the case that journalists, when operating in conflict zones, often embed themselves in Aid/Peacelands and this impacts the types of interactions they have and the potential ways of seeing the world. In international relations this has led scholars to speak of a "media-aid complex" that sponsors and simplifies conflict narratives (Perera 2017; Laudati and Mertens 2019). There is another risk, one related specifically to journalist-source relations: in the eyes of locals, and in particular the most disenfranchised, there is not a significant enough distinction between one westerner seeking to engage with (one could even say professionally use) their suffering, and another; we all exist in the expat world, physically next door but actually occupying a different space-time, making it difficult to truly understand each other. Given the well-documented dispersal between the fields of NGO communications and journalism, including the crossover of professional trajectories, NGO communications officers using the tools of journalism, and journalists accessing interviewees in the company of NGOs (Powers 2016, 2018; Wright 2018), this merging of roles in the eyes of borderland inhabitants is hardly surprising.

I do not believe my interviewees saw me as anything other than a journalist, but what being a "journalist" actually meant for them, and why they engaged in the interviews, is another matter. At times, I felt that my interviewees' primary motivation was to seek material help from the "foreigner," however meager and indirect—which I provided. I paid for taxis that weren't used and I bought sugarcane that I gave away. Looking at the situation as a whole, it makes perfect sense that we are seen, not just as "foreign correspondents" (or perhaps not *only* as), but also as part of an economic system that brings slight material benefit to them and careers to us. Conflict scholar Suda Perera discusses a similar issue in noting attitudes that have become a barrier to peace. Congolese people "feel that conflict has become yet another resource for international actors to exploit for their own gain. As a consequence where locals do engage with international programmes, they often co-opt or adapt projects for their own needs" (Perera 2018: 5). A similar dynamic is reported by Andersson (2020: 29) in Mali, with many locals telling him that peacekeepers were "just there to eat," meaning the peacekeepers were in it only to gain something for themselves rather than help Mali recover from conflict. These same sentiments were also expressed in the borderlands of US-Mexico and Africa-Europe, with informal migrants acutely aware of their monetary value to a variety of legal and illegal actors (Andersson 2018).

This belief that we expats are there to benefit from an economic system of conflict can, in foreign correspondence, manifest in the existence of something called "NGO-speak." I was warned about this phenomenon by a highly regarded local Congolese journalist ("Interview D" 2014) who worked as a stringer for a major European news service. He warned me that, in interviewing local Congolese, I might find their responses distorted by their understanding of how to respond to foreigners who generally represent INGOs. Their responses would not be geared toward a sense of political discourse and engagement in the public sphere, as is usually the case when the "woman on the street" agrees to talk to a journalist in most western countries. Instead, they would be looking to use the interview to seek material help, a response habituated through the dominant pattern of conversation with foreigners.

This warning proved well placed, as I was able to detect this NGO-speak in my interviews, with answers veering off into discussions of potential projects that could be funded by INGOs. I also felt that interviewees' interest in talking to me was strongly influenced by the prospect of material benefit. Indeed I had written about this element of using the interview, or consenting to the interview, in the hope of material betterment, when I was in the Sudanese returnee camp (Giotis 2011). However, I had not previously considered this phenomenon as "NGO-speak." I had thought this type of interaction was just a result of poverty and had not made the connection to the pervasiveness of the INGO experience for people in certain parts of the world. Agier makes the point that adopting humanitarian language is a natural course of events. That the extended experience of being "encamped" brings about contact "with a humanitarian system that is both global and localized" (Agier 2016: 65). Agier also makes the point that refugees are now forced to vie for the position of "most vulnerable." Agier's fieldwork found the category of most vulnerable was used for resettlement, with refugees requiring an extra recommendation from a humanitarian organization before being recommended for the resettlement process (let alone gaining resettlement). In Indonesia this category is similarly used for resettlement, with refugees feeling that they need to be completely broken before they will have their political rights as refugees acknowledged (Giotis 2019b). In my reporting from Goma I found that the category of most vulnerable was required once general food distribution stopped. As noted by Agier (2016: 64), this category only "victimizes the refugee identity still more."

References to NGO-speak can be found directly in development literature and indirectly in journalism literature. For example, development scholars Wallace et al. write: "All across Africa conversations in the vernacular are dotted with English words, which are possibly, by now, universally known to rural communities that have contact with development agencies—gender, strategic planning, rights, advocacy—but which often have little meaning for those using them" (2007: 170). Barry-Shaw, Engler, and Oja Jay (2012) quote a veteran development worker on microfinance as saying that "everyone from camel herders in Mauritania to peasants in rural China can speak the lingo" (2012: 35). Moreover, the particular form NGO-speak takes can be influenced by the media-aid nexus and the repeated storylines attached to specific regions. Reporting from Luvungi, DRC, Laura Heaton quotes an interviewee as describing "systemic rape" and she, and her colleagues, feeling that the interviewees had been "coached" so that the destruction of the village by rebels would fit into the war-rape narrative of the DRC (Heaton 2013). Adoption of the specific rape victim frame can be motivated by the need to fit into existing western priorities in terms of material support. Perera (2017: 10–11) gives the example of a victim of sexual assault being told she should identify the assault as having been committed by a soldier (even through it wasn't) so as to access NGO support. Laudati and Mertens (2019: 74) argue that the strength of the narrative linking conflict minerals to rape "limits women to the position of victim, disregarding the active role women voluntarily seek and play in mining, and in extractive communities in general."

This disturbing, gendered, element in NGO-speak reaches beyond the extreme example of rape. The perceived need to display helplessness has likely created a de facto pattern of engagement between foreigner and local that may recreate images of women as mute victims (Giotis 2019a). In Mugunga camp I was struck by the way one of my interviewees changed demeanor as soon as I signaled the interview was over, showing me for the first time a wide smile and suddenly becoming warm and open. Foreign correspondent Joris Luyendijk (2009) seems to have experienced something similar. Recounting his first visit to a feeding camp in Sudan, he describes conducting interviews in two huts and receiving the same mute response of despair. Finally, as he was urged by a Médecins Sans Frontières official to go into a third hut, in an effort not to succumb to tears, Luyendijk broke from his usual interview routine and greeted his third set of interviewees with an animated "Hello, everybody!" He writes this break from the formal interview structure changed everything:

All of a sudden their faces lit up. Girls giggled, an old man shifted in his seat, and children nudged their mothers. "Look Mummy!" A little toddler of around two wriggled free from his sister, grabbed my knee with both mitts, and tumbled over. Mothers of emaciated infants burst out laughing and used their free hands to wave. (Luyendijk 2009: 3)

This response seems to imply the unusual opening parley of Luyendijk broke the de facto pattern of engagement. Interviewees saw themselves being greeted as humans, and not as humanitarian victims, and so put aside the performative element of their victimhood. The suffering was not less or more between the different huts, only the interaction was different. As Polman notes: "Most refugee camps have television sets that can pick up CNN, so refugees see how 'we' portray victims. They learn to fit the expected image" (2010: 159). What Luyendijk walked into was an example of people engaging in image-NGO-speak, probably with a view to facilitating the transactional frame, and what he ended up experiencing was an affective break from the power of the development discourse.

From a straight reporting viewpoint, how justified is a reporter in including a quotation in a story that may have been contrived to gain material benefit? On one side, you might argue that there is no problem in including it because this manipulation happens all the time, just in different forms. The seminal works of Hall (1978) and Ericson, Baranek, and Chan (1989) have shown the control that sources can exert over the news by the statements they make. Davis and Kent (2013) argue that PR consultants and elites have knowledge of media practices and are able to manipulate the frames journalists use in stories for their own ends.

On the other hand, you might argue that there is a difference between what we, as journalist and as audience, expect (and allow/guard against) from elites versus the "woman on the street" interviews. Perhaps, but this can be problematized further. In a country like Australia there is a belief (albeit held more strongly by some people than others) that we have a civic sphere and our engagement with that public sphere, including through the media, can help shape our society. In a society like Goma, and in particular in the strata of people foreign correspondence mainly engages with, these people's access to power is much more limited. In many facets of life, the state has been replaced by charity (Van Reybrouck 2014: 475) and so engaging in NGO-speak may in fact be engaging with the public sphere. A further complication arises simply because of the power of the development discourse. In my case, my translator, who was also a local journalist and English teacher, had worked in the past with INGOs and wanted to work with them again; he seemed imbued with NGO-speak. His translations of long passages of conversation were often sparse generalities dotted with buzzwords. In this respect hiring a local partner who is concerned with the independence of journalism is important. One of my fixer interviewees specifically mentioned the importance of his work in stopping journalists from succumbing to INGO viewpoint capture ("Interview C" 2014).

Individual levels of poverty also impact these deliberations. The poorer you are, the further you are from centers of power. Are journalists then justified, in their role as watchdogs, in taking up your case, even if it is a direct plea for material benefit? The verdict of my Frame Reflection Interviewees was yes: when people are desperate the journalist should be on their side, highlighting their need for material support. However, discursively, over the long term, danger comes when that storytelling is simplified into advocacy in the transactional-humanitarian frame. People's desperate lives in necropolities should not be divorced *from* politics but tied *to* politics.

All of these considerations need to be taken into account by individual professionals during the crafting of their stories. However, there is also a metaissue, or structural issue, that the profession as a whole needs to acknowledge. In Goma I was also introduced to the term "NGO cement," the common name for the lower-quality cement sold for INGO-funded projects. The fact that people think it's OK to sell lower-quality cement to INGOs tells you something about how INGO work is valued, perhaps because "poor countries are full of [INGO "white elephant"] empty buildings and unused roads. Donor statistics are, too, but under the heading of 'completed projects'" (Polman 2010: 206). If correspondents too are being "sold" our version of NGO cement, what does this say about our value to locals?

One way to consider all of these concerns is to once more go to Harvey's matrix of space and time introduced in chapter 3 and this time center the "NGO-speak" metaphor and connected material, relational, and discursive touchpoints (Table 5.1). This then allows us to examine how the existence of NGO-speak impacts the spatiotemporal relations of foreign correspondence.

Both NGO cement and NGO-speak can be considered metaphors of situated knowledges that, according to the matrix encapsulate both broad social relations of space-time and the way that space is represented and conceived. By recognizing the way in which metaphors of situated knowledge point to existing social relations we cannot hide from the fact that NGO-speak reveals

Table 5.1 NGO-Speak in a Harvey Matrix

	Material space (experienced space)	Representations of space (conceived space)	Spaces of representation (lived space)
Absolute space	Projects built with cement of lower quality sold to NGOs Camp conditions Limited food	Mugunga camp part of UN-INGO world (e.g., World Food Program–branded hessian sacks used to build shelters) Official camp entrance demountable office	Laptop alcove for working in companionable silence at night (Caritas Guesthouse) Refugee camp visit by "Muzungu" (lived space for interviewee) / camp visit by *journalist* (space as lived by me)
Relative space (time)	One-day visit to Mugunga camp Mugunga visit required preapproved paperwork organized for me by local NGO	"Muzungu" "Expat" Expert flying visit Largest UN mission in world Mugunga now "suburb" of Goma	Frustration at "NGO-speak" impacting ability to report, particularly in adding time to writing of story Sense of enhanced understanding after discussions explaining logic of NGO cement
Relational space (time)	My translator, security guard, driver, and guide surrounding me during interviews in Mugunga My translator and his fiancée waiting outside boulangerie with security guard "I Love My Life" bar just outside entrance to Mugunga	Heart of darkness Capital city of East Discourses of development and securitization "NGO cement" and "NGO-speak"	Visions of being in Goma as "courageous" part of work Uncertainty as to how to deal with NGO-speak Desire to ignore existence of NGO-speak because of the way it complicates professionalism

our entanglement in interviewees' eyes with Aid/Peacelands. This is an important recognition. In discussing an ethical framework for media reportage Silverstone (2002: 283) has argued there is "very little sense that we are the objects of the others' gaze, that how *we* are seen and understood by those far removed from us also matters; we need to see and understand that too." The matrix gives a structure to reflexive consideration of how we are seen by our interviewees.

By centering NGO-speak as the focus of the matrix, and then filling in squares in relation to the NGO-speak tension, other elements of Goma space-time also come to the fore—not least it leads to my confession in the bottom-right square that there is a strong desire to simply pretend NGO-speak doesn't exist. At the same time it becomes clear how wrong it would be to do that—for in the end it is the tension that NGO-speak creates that is the truly important point. When Mama Rachel called herself a war victim I struggled with it. As discussed in chapter 2, I worried about the truth of the description. Essentially, I was afraid I was being sold some NGO-speak. But her description was absolutely a true reflection of her local reality— of the ongoing conflict and insecurity that is the norm of new wars. The issue is not the existence of NGO-speak per se, but that journalists fail to give these quotations context. The NGO-speak phenomenon exists, so it is not outside the realm of possibility to contextualize it by paying attention in our reports to the structural forces that have brought these utterances into existence. The point is to move away from seeing static objects of re-portage and instead see subjects, dynamic members of the demos. As noted by Agier (2016: 135), we must focus a "contemporary attention on contexts and processes." In a dynamic world people adopt different descriptions of themselves "depending on the institutional support, the interlocutor, or the chances of access to different resources" (Agier 2016: 135). This sort of nuanced presentation isn't easy, but it could help foster a complex under-standing of the local reality in borderlands sustained by the global "liberal peace" security culture.

The Danger of Us and Them

In December 2020 Reporters Without Borders produced its annual report of dangers associated with news work: 50 journalists dead, 387 journalists detained, 54 journalists held hostage, four missing (RSF 2020a). In some respects these sobering statistics are an improvement. The number of journalists dying, particularly in conflict zones, has decreased since 2015; however, the number of professional and nonprofessional journalists detained is at historically high levels. What 2020 also showed is how quickly the political environment in which journalists operate can change. The pandemic was used around the world as an excuse to significantly step up harassment, especially of local journalists (RSF 2020b).

The particulars of why journalists are targeted can help elucidate contemporary realities of war and global politics. All conflicts also involve an information war, and this used to protect foreign correspondents, who were seen, particularly in nonconventional, rebel-based conflicts, as necessary and neutral messengers (Armoudian 2016: 106). However, as technological developments have allowed more players to produce media, and as the rhetoric of "us and them" has intensified from all sides, the information war has become violent for those now supposedly less needed and less neutral messengers. At times media is targeted to stop reportage. This was seen in the bombing of Al Jazeera's headquarters by the Bush administration (Schwarz 2015). Cottle notes the increased use of "deliberate targeting of journalists (and human rights workers)" by combatants to keep atrocities out of the public eye (Cottle 2009a: 502). Other times attacks are about publicity and broadly shared through social media to dehumanize and feed the "us and them" discourse. This discourse led a military specialist to give the following briefing to journalists heading to Iraq:

> Let's not pretend. After September 11, everything changed. . . . Members of the media—especially journalists from America or Britain—*are seen as ambassadors of their countries*. If you think you've got immunity forget it. Back in the day, people didn't deliberately target the media. . . . They do now. (Ayres 2006: 127; my italics)

The "us and them" discourse, which puts journalists in danger by turning them into "ambassadors," is not limited to opposing sides of a "war on terror"; it is also, more broadly, an economic and emotional discourse. As described by Appadurai:

> A current dynamic of globalization, one noted by a wide range of observers . . . is the growing production of inequality between nations, classes, and regions. This increase in inequality, irrespective of the expert debates about its precise links to open markets and high-velocity global capital flows, is seen at the popular level in many countries as a direct product of the unfettered force of global capitalism and its unquestioned national driver, the United States. (2006: 23)

Later, Appadurai links this popular discourse directly to the *individual bodies* traveling into poorer countries, and links the economic discourse to the military one:

> Americans seem to symbolize the Nikes on their feet and in their missile silos simultaneously . . . and the undeniable Washington imprimatur on some of the most difficult policies imposed by the International Monetary Fund and the World Bank. (2006: 121)

Foreign correspondents, as symbol-wearing "ambassadors," experience the effects of this "us and them" discourse, as do their fixers who come to be associated with them—and this should be related to audiences. In one example, in Pakistan's northern border region, a fixer began to be targeted after civilian casualties caused by US drone strikes. This fixer was kidnapped and tortured and faced threats for years afterward (Baloch and Andresen 2020: 42). There is a type of parallel with peacekeepers. Andersson (2020: 31) states that the archipelago structure of intervention bred resentment in Mali, so that the first target was a popular Bamako expat haunt, the nightclub La Terrasse. However, when frustrations with the international presence in the country led to more and more attacks, it was the African peacekeepers and NGO workers, exposed to the more insecure northern regions, who eventually became the main casualties.

There is a clear ethical obligation to tell the stories of danger faced by our local colleagues—*especially* when those dangers arise because of their association with foreign correspondents. Our use of Peaceland structures likewise increases the ethical obligation to tell the stories of danger faced by peacekeepers due to resentment of the expat-dominated "liberal peace" structure of intervention. Unfortunately, Andersson (2020) notes, in the case of Mali the attacks on local partners received hardly any attention relative to attacks on white expats. Part of the reason for this is existing news hierarchies that deprioritize nonwestern lives based on a calculus of death (Cottle 2009b: 45–50). Here then is another argument as to why this calculus of death should be removed from the profession—it makes the job of explaining borderlands harder. Audiences need to hear more stories of how local dangers are created because of the global "us and them" discourse so as to understand new wars in their global political context.

As with understanding NGO-speak, understanding the "us and them" discourse will require us to pay attention to the way we, the media, are seen. One legacy bequeathed to foreign correspondence by the traditional foreign correspondent is the suspicion of spying. In the late 1970s more than 400 American journalists were in the employ of the CIA (Boyd-Barrett 2004: 38) and, in the 1980s, the editor of "one of Britain's most distinguished journals" believed that more than half of his foreign correspondents were on the MI6 payroll (Keeble 2004: 46).

The idea of the foreign correspondent as spy has not gone away. Picard and Storm (2016: 10) note "a significant challenge for journalists in conflict zones is that some combatants or sides in the conflict may believe the journalists are propagandists or spies for their enemies." In both Sudan and the DRC the "spy" viewpoint had real effects on the way I was seen and the types of conversations I was able to engage in as a journalist. In Sudan, in 2011, a highly educated, well-traveled, and interesting local doctor engaged in conversation with my partner and I during a long-distance bus journey to the Ethiopian border, but at the end of the trip he admitted he held some of his opinions in reserve because he thought we were "probably spies." In the DRC, in 2014, after a few days of using a particular cafe as a spot to meet and interview both western foreign correspondents and Congolese fixers, I was warned that someone from the DRC national intelligence service (Agence Nationale de Renseignements) had set up shop at the cafe table with a laptop to listen in to my conversations. Furthermore, in one of my research interviews, a Congolese fixer told me fixers were always wary of being completely honest with journalists because "next week, after they're gone, you might hear they were from the CIA" ("Interview J" 2014). In the western-spy narrative we again see that there are few distinctions made between foreign correspondents and other international interveners, with research in the DRC finding that "rumours abounded that certain NGOs were government spies, or working for mining companies" (Perera 2018: 21).

More disturbingly, once again the local fixers come to be associated with foreign correspondents—including the accusation of serving foreign agendas. Baloch and Andresen (2020) point out that fixers in Pakistan put themselves in extreme danger to help get news of their local troubles to international audiences, yet correspondents trample roughshod over their glocal reality—for example, ignoring pleas from fixers to present more balanced representations of China's involvement in the region. The price paid for some of the local journalists was accusations of spying and torture and being forced to give up their profession. These are no small matters. Deciding how to deal with them is an ethical minefield that can only be negotiated on a case-by-case basis but, as with NGO-speak, one tactic that may help is openly discussing, in stories, the way we are seen and the complications this creates. Such transparency is necessary to be able to communicate the effects of the "us and them" discourse with integrity. It can also help elucidate the way global currents become violent local realities in borderlands.

#grassrootsglobalization

> The decline in international journalism documented by Utley and others clearly goes to the heart of concerns about an informed citizenry and its capacity for understanding today's global world, its interdependencies and inequalities.
> —Cottle 2008: 347

I wish to make a bold normative proposition. Empire and colonization were globalization from above. As noted in the introduction, foreign correspondence played its part in this type of globalization through the prototype foreign correspondent, Henry Morton Stanley. Therefore, I would like to argue that part of the decolonization of the profession is to contribute to globalization from below—grassroots globalization controlled by the demos via broad-based cooperation. In their discussion of the urban war-zones that have become a feature of new wars, Kaldor and Sassen ask for more attention to be paid to everyday strategies of resistance found in strategies of survival.

> Such minor and even major survival interventions by the women, men and children living in cities at war are not often described in analyses of war situations. We have found that it might be worthwhile to consider such interventions, minor as they might be. Rather than seeing these modest efforts as somewhat irrelevant to the analysis of cities at war, we join the growing movement to recover the daily lives of people struggling in a war situation. (Kaldor and Sassen 2020: 228–29)

This visibility of daily resistance is an important point, but journalists undertaking foreign correspondence need to be clear that simple stories of victimhood and survival are not enough. Simple media depictions of war and catastrophe only work to inoculate audiences in the west against engaging with the risks of a globalized world. "Immunization" is "achieved through measured doses of the same distress-provoking stimuli" (Meek 2016: 3). This protects the psyche and enforces the separation between them and us, between here and there. This is not a state of affairs that can continue.

The need to find reporting methods that break past these numbing effects, and instead foster understanding of the interdependence and inequalities of our shared world of risks, has become even more urgent since 2020. The spread of Covid-19 highlighted how interdependent our world truly is. It also highlighted, yet again, the role of borders in exposing power relations—with vaccines literally being stopped from transport on airport tarmacs. The Covid-19 crisis, like other global crises, created and reinforced inequality, leading the IMF to warn of a "great divergence" postpandemic as the poorest take longest to recover and inequality grows (Georgieva 2021).

This chapter has made the case that borderlands need to be understood not as places of chaos "over there" but as spaces subject to *glocal* forces. We must also remember these forces are not a one-way street but instead influence our entire globalized world. In a discussion of the necessary conditions for grassroots globalization, economics professor Jayati Ghosh highlighted that global processes of capital not only harmed poor southern countries but also poor communities in the north ("PS Events" 2021: @35m). She spoke of the need for international solidarity but feared it was a pipe dream because the western media were failing to explain global interdependencies.

By contrast one area where western media regularly describe global interdependencies is financial markets news. Berglez (2013) points to economic news as an example of what is possible in terms of the national media's role in fostering a global news culture, and thus global political culture. At the same time Berglez (2013: 71) astutely notes that there is an elitism inherent in this type of journalism that causes divisions and works against "global empowerment of the majority population." Berglez argues:

An important aspect of [economic] news is that the global dimension of society *is there all the time* in terms of a regular flow of global outlooks on the economy. For the financial elite, the global condition is rather [a] "normal" dimension of society and an integrated part of life. For the non-elite, on the other hand, who in most cases does not consume this kind of news, the constant news flow of economic figures from the world's stock markets and the related political conversation might cause alienation and give them the feeling that the global world is for someone else (that is only for the successful class) and/or that this reality is impossible to affect or move in a more democratic direction. (2013: 69–70)

As traditional foreign correspondence has shrunk, we have, conversely, seen the growth of premium news service correspondents identified by Hamilton and Jenner (2004) over a decade ago, and this, combined with strong strategic planning by more broad-based but still financial-specialist media, means that the financial journalism sector is investing in understanding the world. In many respects this is good news. After all, who better than a meld of financial and foreign correspondent to explain the significance, including its impact on people, of New Orleans in reconstruction being termed "Baghdad under water," complete with Halliburton's dominance of reconstruction contracts and a missing $12 billion (Polman 2010: 199)—a perfect example of Harvey's point (2006: 12) regarding experimentation at the periphery (Iraq) being moved to the center (United States) or perhaps an example that peripheries and centers are no longer aligned to states but to enclaves of capital. However, this foreign correspondence cannot remain exclusively in the domain of premium service news.

Journalism has always claimed special privileges in the political and cultural sphere through its commitment to serve "the public"—broadly defined. The foreign correspondence needed today should aim to serve a global, inclusive, public sphere, for any form of globalized endeavor that involves divisions and inequality holds the potential for disaster. This danger is seen in the security culture of liberal peace that, through its divisions and insecurity, holds within it the fertilization of identity politics (Kaldor and Sassen 2020: 13). These seeds spread beyond borderlands. In western countries identity politics are fueled by fear of the other who is being expelled by forces of conflict and exploitation. The sense of impotence in the face of global processes leads to xenophobia, right-wing nationalism, and rejection of national democracy and international institutions.

These forces of deglobalization and de-democratization have coalesced with a crisis in trust. For years the global Edelman Trust Barometer has been noting a steady decline in trust in institutions. At the same time there is widening trust inequality. In Australia in 2020 trust inequality was the highest in the world ("Edelman Trust Barometer" 2020). That is, higher-educated, wealthier people trust institutions, while poorer ones don't. This divide is true of trust in the media too (Park et al. 2021: 76); in fact the media may be the most crucial piece of the puzzle. Blöbaum (2014: 22–26) argues that trust in institutions is created through trust in journalism and trust in journalism is created through trust in its processes. If that is true, the responsibility on foreign correspondence to help rebuild global trust is huge, and that trust can only exist if global journalism is seen to hold true to its community-building and watchdog processes. In its community-building role foreign correspondence must undertake to explain global interdependence as a shared present. In watchdog mode it is necessary to take very seriously the possibility that the "liberal peace" in fact "creates an industry of insecurity that may self-perpetuate and add to existing insecurities" (Kaldor and Sassen 2020: 13).

This trust will not be easily gained. In the Frame Reflection Interviews I conducted my interlocutors did not trust that journalists traveling to borderlands were able to uphold their professionalism as watchdogs. They felt that a lack of preparation, time, self-censorship, and the influence of powerful politico-economic interests were insurmountable barriers when it came to complex reporting. They also discussed lack of trust in a way that pointed to an us/them divide, for example when Aristotle[2] said, "Wars start in Western countries and are finished in Africa," or when Graham pointed to the fact that correspondents could do harm to individual refugees if they quoted them as being critical of how the system works. Moreover, both Graham and Maria assumed that foreign correspondents did not, or could not, understand the precarity of lives in borderlands and so interviewees had to be wary of them. Perhaps the only way to overcome this is to build up a body of complex reporting on borderlands. Only then can people be sure that foreign correspondence is there as part of a service to the global public sphere instead of simply a career opportunity.

The lack of trust described by my interlocutors is natural as the current situation in borderlands speaks to an engrained system that did not spring up overnight. When investigative journalist Antony Loewenstein set about to document the "vulture capitalism" benefiting from privatized wars and development in postcolonial countries, he found his belief in journalism and its role in reporting countries devastated by globalized, corporatized conflict "fundamentally challenged" (Loewenstein 2013: 212). Loewenstein came to believe that the media elite joined the business and political elite in silencing the majority while claiming to represent them (2013: 214). However, ultimately, he came back to journalism as a tool that could tell important stories of the reality of the world today. Journalism can and does.

Looking at the feature articles written from Goma (described in chapter 2), those pieces of journalism clearly contribute to this endeavor of recovering grassroots resistance to conflict—of making it visible, with all the possibilities that entails. Yet I cannot pretend to have done so consciously. Nor can I pretend to have developed, at that stage, the proper appreciation for the *glocal* nature of borderlands and the need to report them as such. Thanks to the influence of the FRIs, the local-global aspect is there, to some degree, but enough knowledge and determination to write at the glocal nexus were not. This determination will be found by decolonizing foreign correspondence and reimagining it as a professional practice of normative cosmopolitanism.

6

Decolonizing and Reimagining

An authentic professionalism (as opposed to one manufactured for the purpose of media image management) may be one of the most important counterweights to the economics of global newsgathering.

—Reese 2001: 233

Dateline: London, England, July 17, 2014

Standing in Stanford's—that bastion of English life—first introduced to my consciousness through Arthur Conan Doyle's rendition of Edwardian London in *The Hound of the Baskervilles*. This is *the* place for maps and travel research. During my years living in London, before every trip, it was a necessary part of my life. Now, here again, on a whistle-stop tour to find books on the DRC, books available in this basement in Covent Garden but not in bookshops in Australia. This isn't surprising, it is evolution. Australia has its own colonial history but not as a metropole where knowledges are stored and used as power.[1]

One of those books is *Radio Congo: Signals of Hope from Africa's Deadliest War* (Rawlence 2012). I flip to the blurb on the back and am immediately struck by the reviews. Adam Hochschild, author of *King Leopold's Ghost*, writes that Rawlence is "undaunted by hazards that would send most journalists scurrying back to the bar of their hotels." *The Times* reviewer notes: "It is his closeness to the Congolese that really stands out."

Consider the critique implicit in these comments; foreign correspondents in the Anglo-American journalism tradition are seen as hiding out in hotel bars and removed from ordinary Congolese, and this is an *understood* critique, something it is assumed readers in bookstores will identify with. We have to ask ourselves, do we have an authentic professionalism based on values, or, like the theater of datelines, discussed in chapter 4, is it all about media image management? To justify its place as a meaning-making practice

Borderland. Chrisanthi Giotis, Oxford University Press. © Oxford University Press 2022.
DOI: 10.1093/oso/9780197565797.003.0007

in today's world, foreign correspondence has to be a profession that speaks to values. This requires a decolonization and then a reimagining.

This final chapter will consider the epistemic basis for foreign correspondence. In philosophy epistemology is understood as the theory of knowledge and driven by three questions: "What is knowledge?" "What can we know?" And more recently, "How do we know what we do know?" (Greco 1999). For journalism Ekström (2002) argues that we can consider our institutional epistemology as formed via a tri-part relationship around the form of knowledge (e.g., genre of the report; TV/print, investigative/news), the production of knowledge (e.g., routines, methods, and practices), and the public acceptance of knowledge claims. Thus, if we apply the foundational three questions, "What is knowledge? What can we know? And how do we know what we do know?" then we can see that foreign correspondence would benefit from decolonization processes across the three areas raised by Ekström. The discussion of tropes and frames in chapter 1 showed how the genre of foreign correspondence needs to evolve to excise the false knowledge of colonialist tropes and explain new glocal postcolonial realities. The routines and practices of foreign correspondence likewise need to change so as to access a broader range of discourses outside the Aid/Peaceland boundaries that will lead to new knowledge. However, it is the third element of the epistemology that is perhaps the most important to consider.

Ekström (2002) argues that there is an essential knowledge claim common to all kinds of journalism and crucial for its public acceptance as a form of knowledge. This is "its claim to present, on a regular basis, reliable, neutral and current factual information that is important and valuable for citizens in a democracy" (2002: 274). This argument is perfectly consistent with the widespread belief that journalism acts as a fourth estate and also extends the eyes and ears of the public so that modern democracies can function with citizen participation (Wyatt 2010). This epistemic basis also places foreign correspondence in a particularly vulnerable position, for who is this citizen to whom foreign correspondence knowledge claims are aimed? Historically foreign correspondence was aimed at the citizen of the empire, and the knowledge that was produced followed suit. Without a clear sense of the new global citizen, one that instead embraces normative cosmopolitanism in a world-risk society, the professional knowledge production of foreign correspondence is on shaky ground. The lack of an imagined global citizen goes to the very foundation of our epistemology, for what is *journalistic* knowledge production without that citizen?

This shaky epistemology is the reason, I believe, that foreign correspondence in borderlands is too often captured by the concerns of the powerful neighboring field of international development. Yet, ironically, borderlands are exactly the place where this new cosmopolitanism will be found, for borderlands enforce recognition of our connected world—and foreign correspondents are key professionals in discovering this knowledge. So let me take you once more to Goma, to the hotels, and to Mugunga refugee camp and now, instead of considering these spaces as an Aid/Peaceland in which foreign correspondents become *embedded* and our professionalism dissipated, let's consider what would happen with a strengthened professionalism, foreign correspondents with a strong sense of the value they provide through the glocal knowledge they create—could we then find ourselves existing in a heterotopia?

The Glitter and Clash of Knowledge Creation

The concept of heterotopia to which I refer comes from Foucault. In its Greek root it means "different place," and thinking about heterotopias as a place housing *difference* is a useful start. Robert Topinka notes that most scholars discuss heterotopias as sites of resistance, but his interpretation focuses instead on the way they create "an intensification of knowledge" (2010: 55): "Juxtaposing and combining many spaces in one site, heterotopias problematize received knowledge by revealing and destabilizing the ground, or operating table, on which knowledge is built" (2010: 56). He further describes heterotopias as places that are both open and closed, "penetrable" places (2010: 57) defined by the paradox in which they are both

> separate from and yet connected to all other spaces. This connectedness is precisely what builds contestation into heterotopias. . . . Instead of remaining always separate [as utopias], heterotopias hold up an alternate order to the dominant order, providing glimpses of the governing principles of order . . . Heterotopias reconstitute knowledge, presenting a view of its structural formation that might not otherwise be possible. (Topinka 2010: 60)

I argue that the borderland of Goma could be a heterotopia. Certainly, there is the juxtaposition and combination of many spaces in one site. The hotels, coffee shops, and clubs of the Aid/Peaceland element of Goma serve as offices, homes, sanctuaries, debating venues, fun hook-up joints, spaces of angst and guilt, and sites where the local population can connect to a planetary order of "development."

The borderland of Goma also meets the definition of being separate but connected; it is absolutely its own space, distinct in the world, yet it is intimately connected to a larger planetary structure of governance, as it creates and is created by the Aid/Peaceland it houses. The Aid/Peaceland is separate from yet connected to Mugunga refugee camp, which is separate from and connected to the city of Goma. Mugunga, in turn, through its representation as a refugee camp, is connected to Canberra politicians and their re-election slogans (to take just one example), while its actual functioning through the global funding budget for food is connected to conflicts all over the world jostling for budgetary priority. The DRC intervention and the specific interveners are distantly and yet unequivocally connected to the IMF and World Bank in New York. Mugunga and the Aid/Peaceland spaces also exist because of and in concert with the militias, corrupt politicians, and business middlemen who trade conflict minerals and the multinational UN peacekeeping mission tasked with disrupting this trade. And, because legal and illegal trade does happen, these spaces are connected to stock exchanges around the world.

The point is, all of these glocal connections, and more, are there to be seen.

A heterotopia is further defined as a "space in which we are 'drawn outside of ourselves' and 'the erosion of our life' takes place" (Johnson 2006: 78). Referring to this element of "drawing out," Johnson notes that it is "crucial. Heterotopias draw us out of ourselves in peculiar ways; they display and inaugurate a difference and challenge the space in which we may feel at home. These emplacements exist out of step and meddle with our sense of interiority" (Johnson 2006: 84). I saw this "drawing out" happen repeatedly in Goma, not in the official workspaces of Aid/Peaceland but in the interstitial spaces of clubs, bars, cafes, hotels, parties hosted at share-houses and hotel restaurants—it was in these places that questions regarding people's modes of practice and the very justification for their profession were raised. Witness too, the point made by one of my stringer/fixer interviewees, that foreign correspondents do acknowledge, *in their private conversations in the interstitial Aid/Peacelands of Goma*, that their storylines are flawed (2014).

Obviously these emplacements are meddling with our sense of the world and of ourselves in it. Agier argues that one of the functions of a heterotopia is "epistemological decentering" (2019: 15) but does this meddling with our interiority also result in the "intensification of knowledge" described by Topinka (2010: 55)? Are we truly "revealing and destabilising" received knowledge (2010: 56)? The answer, if we look at the flawed storylines being admitted to, is no. Heterotopias produce knowledge when "epistemes collide and overlap" (Topinka 2010: 55)—it is a process. Topinka notes that "for Foucault and Nietzsche, knowledge does not exist except as a product of combat" (2010: 64) and later adds that "Nietzsche describes the battle that produces knowledge, and Foucault renders that battle in space. Heterotopias, with their intrinsic contestation of order, are spatial organs of knowledge production" (2010: 66). Thus, at least one element that is missing and which stops foreign correspondents from finding borderlands a heterotopia, is battle, and it is missing because of a weak sense of foreign correspondent knowledge claims.

This weak sense of epistemology is why the transactional frame provides a sense of "doing something useful" when confronted by tragedy. Polman has described how, "confronted with humanitarian disasters, journalists who usually like to present themselves as objective outsiders suddenly become disciples of aid workers . . . elevating the trustworthiness and expertise of aid workers above journalistic skepticism" (2010: 47). This subordination of our professionalism is poignantly shown when *aid workers themselves* are looking for more professional pride from journalists:

> Do aid workers use journalists? Of course they do, said Jacques de Milliano, former director of Dutch MSF. "To raise funds. It's the job of journalists to provide balanced reporting, to refuse to prostitute themselves to aid organizations. There ought to be an element of journalistic pride." (Polman 2010: 46)

In subsuming our professional identity, foreign correspondence becomes captured by Aid/Peacelands, and the possibility of somewhere like Goma acting as a heterotopia dissipates.

Another element that may be currently missing in constituting Goma as a heterotopia is penetrability. The Aid/Peacelands of Goma are penetrable, as seen by my easy entrée to them, but I am relatively/relationally "white." The impenetrability for others was brought home to me through my translator. I did not find my translator through the usual methods of Aid/Peaceland, but instead through a connection from the east DRC community in Sydney.

When I asked him to meet me at the Boulangerie, which he had never been to before, he did not go inside but instead waited outside with the security guard until I arrived—for my translator, who was not part of the comprador class, the interstitial space of the Boulangerie did not feel penetrable.

Greater securitization seems to be the way Aid/Peacelands are heading (Duffield 2012). This not only positions the local (insecure) space as impenetrable by interveners, but will also lead to lost access opportunities for those already struggling to be heard. No longer will a situation like the one described by Lederman be possible. Writing of the first Intifada he says, "The Palestinians, because they had no official spokesmen, had to rely almost entirely on informal contact with the press—whether in the field or at some central location. They had an ideal venue—the American Colony Hotel in East Jerusalem" (1992: 153). That sort of scenario is disappearing, yet a solution is at hand. One of the great joys of being a journalist is, metaphorically speaking, our "access all areas pass"; armed only with our notebook and our professional identity (and in borderlands add a fixer), every echelon of society is open to us. But it's not easy. There is no point denying increased risks to journalists and increased targeting. Nevertheless, devaluing the profession through separation cannot be the long-term solution, for, like the archipelago peacebuilding interventions discussed in the previous chapter, there is a contradiction between ends and means. On normal reporting assignments in borderlands it is not beyond our capabilities to expand *our* Goma beyond the security fences. We can choose to stay in the hotel *without* the running water, the one that is open to the community, and only visit the other hotels, or we can choose to mix and match. Personally, I would recommend not overvaluing running water, as sharing some minor daily battles and staying in places open to the community cultivates the affect of solidarity. Either way, consciously spending time outside the fences turns this expanded space of our Goma into a valuable heterotopia, maintaining penetrability and connections between different spaces—indeed we become one of those connections. In my social conversations I transferred knowledge of Goma outside the security fences of its Aid/Peaceland to those inside the fences.

This conscious use of our "access all areas pass" aligns with our already existing professional identity—with senior journalists arguing that foreign correspondence is at its best when it is based on "shoe-leather reporting"[2] (Sambrook 2010: 37; Hannerz 2004: 163). Building on this professional conception of ourselves as seekers of truth by getting out on the street, we thus strengthen our *unique* professional identity, and this will again help create the clash of episteme needed for knowledge-creating heterotopias.

Another argument for the cultivation of heterotopic space comes from Zelizer's (2017) provocation—that strengthening and expanding the professional contours of journalism requires a capacity for imagination, as indeed does the practice of journalism itself. Imagination is nurtured in heterotopias:

> As with Lefebvre's notion of heterotopy, there is here a teasing play of differences, but without the logic of contradiction, or negative dialectic, or any consequent utopic formulation. They offer no resolution or consolation, but disrupt and test our customary notions of ourselves . . . With different degrees of relational intensity, heterotopias glitter and clash in their incongruous variety, illuminating a passage for our imagination. (Johnson 2006: 87)

Consider too the words of Van Reybrouck, who, in the acknowledgments section of his history on the Congo, gives credit to the neighborhood in which he wrote it. He says that the neighborhood helped to shape the critically acclaimed work:

> I wrote *Congo: A History* at my studio in Kuregem, the oft-cited "problem neighbourhood" in the Brussels district of Anderlecht . . . where I have worked with pleasure for more than four years. I could never have dreamed of finding in Europe a better place to write about Congo: my studio has a view of the street where every day dozens of second-hand cars are bought and sold before being shipped to Central Africa. The street corners are adorned with posters for concerts by Werrason or services by faith healers. From the outside this neighborhood seems so poorly integrated into Belgian society, I am sometimes told, but from here it seems more like Belgium is poorly integrated into the world. Kuregem is a lesson in globalization, but also in empathy and involvement. (2014: 559)

I do not believe it is a coincidence that Van Reybrouck's feelings about Kuregem mirror my feelings of indebtedness during this research to my upbringing in a "problem neighborhood." Such neighborhoods create the "relational intensities" described as necessary for heterotopias, and the degree of intensity is created by the density of diversity. Van Reybrouck wrote his acknowledgments in 2010;[3] by 2016 Kuregem was no longer being studied as a "problem neighbourhood" but instead as a successful neighborhood with lessons the World Bank is keen to learn and replicate. One of the keys to its success is considered to be the density of population (Saunders 2016). My neighborhood too has changed; its "problem" past has faded into myth through gentrification. I don't miss the drug dealing and robberies, but I do miss the density and variety of the population. Large residences once partitioned into flats and boarding houses have been turned back into grand homes, and one small family now lives in the space previously occupied by several, often of various nationalities. There is less opportunity for the "glitter and clash" of "incongruous variety"; there are still people of all walks of life, but not in as high a density. The affects created by my passage through the neighborhood have changed, and my imagination is not *constantly* challenged as it once was. That is why, as part of the Frame Reflection Interviews, it was important to *spend space-time* in the Congolese church and space-time hanging out at the fantastic multicultural hub that is Liverpool Library. It is this sitting in heterotopic space that opens the imagination and catalyzes successful frame and discourse diversity.

Does this mean that we can dispense with the FRIs described in chapter 2 and simply hang out in culturally heterotopic spaces? No—it is one thing to be open to knowledge and another to actually possess that knowledge and then to act on it in practice through writing different stories. Furthermore, in a profession like journalism that is reliant on text and narratives, we must consider that Foucault discovered heterotopias, not only in physical spaces but also in the clash of words as cultural codes. Foucault's textual heterotopias, again found through juxtaposition, are "unthinkable spaces that reveal the limits of our language [and] splinter the familiar" (Johnson 2006: 85). The revelation of an unthinkable space is exactly what I felt I discovered through my FRIs. The familiarity of what I thought I knew was splintered when my interviewees in Australia showed me a different way of thinking about the victim frame, a way that did not nullify the agency of the people involved. This happened again when my contacts in Goma showed me a different way of thinking about the conflict minerals frame. I felt the limits of our current cultural codes in struggling to write articles portraying both victims and agentful human beings, and in imagining a way past these limits it was for heterotopic openings I searched.

Van Reybrouck too, did not limit his heterotopic engagement to absorbing the affects of his neighborhood; he also actively worked to discover the "glitter and clash" of narratives in discourse. I know this because another space that finds its way into his acknowledgments is the Royal Flemish Theatre in Brussels—the site of a long-term exchange program between Congolese and Belgian artists that started at the same time as Van Reybrouck started his research, and which he was involved in from the beginning (2014: 559–60). Like the FRIs, what Van Reybrouck seems to be describing involves partnership with others and knowledge creation in a situation related to, but separate from, the actual content of his book.

Of course the original FRIs I conducted in Sydney were not just about consciousness, they were about affect and openness. This sense of openness, generated by my positive experiences while conducting the FRIs, was reinforced by my experiences leading up to my entry to Goma, starting at the DRC embassy in Athens. The staff there not only went out of their way to help me acquire my visa, but also gave me a lift back into the center of Athens! They were carpooling and saw me walking down the road and stopped to pick me up. I happened to be on the phone to my mum at that moment, and one of the embassy workers said to me: "It's good we've picked you up so you can tell your mum there are good people in the Congo too. There are bad people, but there are good people too." The statement stayed with me for the fact she was quite obviously working to destabilize knowledge of the DRC not only for me, but for my mother, who she knew would be worrying about me.

I chose not to enter Goma by a direct flight. I chose to work my way to Goma slowly and gently. I stopped off for two nights in Egypt, visiting friends and then for a night in Rwanda, again visiting friends. Because of the down-to-earth nature of my friends in Rwanda, I realized that my university's insistence that I take a private car from Kigali to Gisyeni was an unnecessary capitulation to the biopower of securitization. I could easily take a local bus. As always, I had fun on the local bus, talking to people and learning as much as possible. Once off the bus, I simply followed the flow of people who were changing sim cards and walked to the "little border" used by locals. All of these things contributed to a relaxed and happy attitude and, I am sure, led to my extremely easy visa formalities. Indeed, I later found out that my driver and fixer for the trip to Mugunga had been waiting for me, against my wishes, at the "big border" crossing usually used by foreigners. My entry to Goma would most definitely have been different at the big border crossing—it would have been structured around my whiteness. This doesn't mean that an open and unstereotypical experience was not possible at the big border. It would have depended entirely on the mix of people, their former experiences, and their mood at the time—affect can always escape attempts to control it (Anderson 2012). However, by asserting my independence from Aid/Peaceland, I was also hoping to underscore that I didn't want to be funneled into a well-worn groove and, as a journalist, get the same old story—I was looking for something different. I can't be sure that this strategy worked, but my transgression of the spatial boundaries of securitization was certainly something that both expats and compradors remarked upon constantly.

I was able to hang on to elements of my initial open and relaxed attitude for much of my time in Goma, but the more I interacted with the expat community, the more I accepted a competing force—the securitization of space. As a lived space, Goma was being re-presented to me as a place of fear where expat life must be protected physically and emotionally via high walls, private transport, and intense socialization with other expats and select compradors. After leaving Hotel Versailles, I found myself more often concerned with my safety, even though I was now in a space considered safer. I believe that my growing concern was fed by the constant safety talk and by the safety structures visible everywhere. I was also more closed off emotionally and much less often in nonexpat spaces—and I grew comfortable with that fact.

Duffield has described what I experienced as a "deepening paradox" in the aid industry. That is, while the industry is determined to stay in areas increasingly perceived to be dangerous, "The actuality of aid bunkerization and aidworker anxiety . . . reveals . . . apparent expansion of the aid industry [and] simultaneous social, intellectual and emotional withdrawal—a growing *remoteness*—of international aid workers from the societies in which they work" (2012: 478). I found this remoteness grew in me through the way Goma was re-presented while I was there, by other expats in the Aid/Peaceland in which I was embedded, and through the affect generated by the material experience of bunkered life.

Could I have reactivated openness through different experiences? I believe yes *if* I had been conscious enough to recognize what was happening, but that consciousness had not fully developed. What was missing was an understanding that actually, as foreign correspondents, part of our professional identity is being aware of the space-times we are inhabiting, and this should include using and embracing the heterotopic affordances of borderlands.

Turning heterotopic space into one of the tools of our trade can only strengthen the epistemology of foreign correspondence. In cultivating glocal understandings we create unique value, in heterotopic engagements knowledge and imagination are nurtured, and in narrating the glitter and clash of these spaces the strands of debate that are necessary for a global public sphere to operate come together. In opening the space for empathic discourse, that elusive global citizen too comes into being.

However, as stated at the beginning, for heterotopias to truly exist we need to have a strong sense of the knowledge-construction foundations of our profession. What is needed is a decolonized and expanded understanding of the principles of our professionalism. It is now time to re-examine the idea of "bearing witness."

Beginning Again at Bearing Witness

> Anti-racism requires historical memory rather than an erasure of historical consciousness and simply forgetting, but it also requires relating contemporary conditions to historical conditions and relating local to global. . . This also incorporates a move toward disciplinary decolonization and a more detailed exposition of the colonialist heritage of each discipline as well as an understanding of the historical construction of racist images and conceptions of an objectified other.
>
> —Power 2006: 36

Traditionally, the existence of foreign correspondents has been justified by the need to bear witness, to see and report that which must not be kept quiet and which it is often easier for an outsider to report. For example, one of the UK's most respected senior journalists, Jonathan Dimbleby, once broke news of a famine that Ethiopians living under the rule of Emperor Haile Selassie didn't know about:

> It was a curious irony of modern mass communication that Ethiopians in Addis Ababa thus learned about the starvation that was taking place in provinces just a few hundred kilometres to their north by means of second hand information from those of their compatriots who had seen Dimbleby's [1973] film in Europe. (Hancock in Harrison and Palmer 1986: 59)

This is not a one-way street with western foreign correspondents taking up a white man's burden. For example, journalists in Indonesia have exposed issues concerning the Australian government's barbaric handling of refugee movements that journalists in Australia were unable to access.[4] There is also growing recognition of the importance of transnational collaboration, as with the International Consortium of Investigative Journalists, a group of 220 journalists around the world who work together on major stories of global complexity such as offshore havens used for tax avoidance (Ryle 2016).

My interviewees in Goma repeatedly referred to bearing witness without using the term itself. When I asked fixers whether the western foreign correspondent should just cease to exist, the answer was along the lines of "No, because we need you to report what we can't" (Tapsell 2014; "Interview H" 2014; "Interview L" 2014). That is, in Goma the foreign correspondent is able to file reports that involve great courage and the willingness to take on high levels of risk, but if a local journalist were to do the same, the risks would be even greater.

Despite the fact there is nothing inherently colonizing about this process, this interplay with risk is where the idea of bearing witness picks up colonial echoes. This acknowledgment must become part of our professional historical memory so as to decolonize bearing witness. My argument follows in the footsteps of Power (2006), who calls for disciplinary decolonization for the development profession and who in turn was building on an article by Peake and Kobayashi (2002) arguing for the disciplinary decolonization of the geography profession. Their argument is this can only be done through "an explicit effort . . . to address and correct the various (and often hidden) racist practices and discourses that permeate the epistemological foundations . . . and the institutional structures and practices that shape our work environment" (2002: 50).

My contribution to decolonization in this book has been a detailed exposition of the colonialist heritage in writing tropes and a consideration of journalism practices in borderlands. Decolonization is the necessary start to a strengthened concept of bearing witness. Without decolonization, we are building on rotten foundations. The next step is to clarify exactly what the term "bearing witness" can mean when building value for the type of knowledge created through foreign correspondence.

The professional basis for bearing witness is found simply in the act of being there to report what must be known. But it is more than that—it is also a pact of trust with the audience embedded more broadly in the foundational concepts of journalism. Allan and Zelizer, speaking particularly of war correspondents, have said:

> The act of witnessing, of seeing for oneself . . . forms the basis of the journalist's relationship with diverse publics. . . . Truth-telling, it needs to be acknowledged, is necessarily embedded in a cultural politics of legitimacy; its authority resting on presence, on the moral duty to bear witness by being there. (2004: 5)

Being physically present also provides storytelling capabilities in a way that journalism of assemblage as a "foreign affairs correspondent" cannot.[5] It is no accident that foreign correspondents believe some of the most "inspired" writing in the profession gets done in hotel rooms (Hannerz 2004: 134). There is a physical power in immediacy; this is why the term *bearing witness* is used as opposed to simply reporting. The correspondent must physically and emotionally experience the horror and pain of the situation and transcribe that in the witnessing. This is a powerful affect and an important ingredient in truly communicating the situation to the audience—turning them into witnesses too. Moreover, correspondent Max Stahl has argued "the fact that you are there at all and able to film the story is a kind of hope" in humanity (Stahl in Leith 2004: 349). The foreign correspondent is there too; thus the biopolitical separation of humanity is not complete and this too is transferred to the audience.

This moment of bearing witness is not limited to crisis and death; it relates, too, to life. Philosopher Alain de Botton (2014) makes an interesting argument for the importance of ordinary human moments. He points to statistics showing that on the same day the BBC website recorded 2.52 million readers for "Bowie Comeback Makes Top Ten Singles Chart," it recorded only 4,450 readers for "East DR Congo Faces Catastrophic Humanitarian Crisis" (de Botton 2014: 80). The reason de Botton supplies is that the term "crisis" cannot create tragedy because it is the only way in which the DRC is understood:

> We don't know whether anyone has ever had a normal day in the Democratic Republic of Congo, for no such thing has ever been recorded by a western news organization . . . in truth, we can't much care about dreadful incidents unless we've first been introduced to behaviours and attitudes with which we can identify; until we have been acquainted with the sorts of mundane moments and details that belong to all of humanity. A focus on these does not in any way distract from "serious" news; it instead *provides the bedrock* upon which all sincere interest in appalling and disruptive events must rest. (2014: 84; my italics)

Likewise, when Hannerz asked the foreign editor of *Dagens Nyheter*, Sweden's largest morning paper, how foreign correspondence can create a connected, welcoming global community, the solution the editor offered centered on "more reporting that portrayed everyday life elsewhere" (Hannerz 2004: 27).

Discovering everyday interactions could help to powerfully convey the reality of borderlands. Nadine Gordimer argues that to avoid the accusation of stealing the drama of black lives for their own benefit, white writers must fulfill the essential gesture as described by Chekhov; that is, they must "describe a situation so truthfully that the reader can no longer evade it" (Gordimer 1985: 151). Moeller similarly notes: "The responsibility to tell stories is not a light one—it does not begin or end with a facile relating of information" (2004: 73), or, as Aristotle put it back in antiquity, it is the storytelling that allows audiences to turn representations of horror into understandings of tragedy (de Botton 2014: 193). Applying this distinction to reporting of the east of the DRC, we find there is a lack of narrative around the horrific facts being presented, and so the audience cannot feel the tragedy involved. Similarly, Stahl argues there is great danger in grabbing attention through horror without giving the audience something to share in the story: "If you simply grab their attention over and over again you are depriving and impoverishing the audience and prostituting the image" (Leith 2004: 344).

This difference between bearing witness (narrating tragedy) and testifying (affirming the fact of specific horrors) is a boundary marker between foreign correspondents and humanitarians (Tumber 2013: 62). Yet recently, the idea has emerged that NGO communications people can be the "boots on the ground" (Powers 2016) for the foreign correspondent (or foreign affairs correspondent). This substitution devalues the role of bearing witness. Without physical presence, we blur the position of our legitimacy with the audience and we deprive ourselves of the *ethical act* of physically bearing that witnessing in the body of the correspondent.

Furthermore, the acceptance of NGOs as "boots on the ground" weakens the profession through abandoning our fourth-estate role in a global public sphere. Powers (2016: 14) points out that "NGOs are large organizations with substantial budgetary needs. If journalism is about both the provision of information and the holding of power to account, then the rise of NGO information work raises questions about who will hold NGOs accountable." Accepting the information supplied by INGOs as the raw material for our stories weakens our ability to promote frame and discourse diversity. As Powers puts it: "The broad public acceptance of human rights should not detract from the fact that human rights remains a perspective—that is, one way among many others of interpreting events" (2016: 14).

NGO substitution involves a physical detachment from being there. However, at the extreme other end of the scale, the value of bearing witness also needs to be reaffirmed in relation to what has been termed "the journalism of attachment." Proponents champion this journalism in opposition to objectivity as "journalism 'which cares as well as knows'" and which involves evincing an emotional attachment to the "good guys" in any conflict (O'Neill 2012: online). For its detractors it is "highly selective" (Tumber 2013: 65), "opening the door to mistaken accounts of conflicts" (Tumber 2004: 201), and "overlooking complexities and political nuances . . . substituting morality tales for tough reporting" (O'Neill 2012: online). Journalism of attachment could also be seen as playing into the dangerous "us and them" discourse at a time when independent information is more important than ever.

> Today, modern conflicts can be as much for public opinion and the politics of identity as they can be for geographic territory. This places journalists and media in the front line. It exacerbates the mindset of "you're either with us or against us," which delegitimises any independent of neutral role. Yet it also increases the importance of available, independent non-partisan information. (Cottle, Sambrook, and Mosdell 2016: 201)

I am not unsympathetic to the genesis and aims of the journalism of attachment. As with the transactional frame, it comes from the horror of the situations journalists face, and, as described by one of its key proponents, the need to do "more than journalism" (O'Neill 2012: online). However, as with the transactional frame, it is this idea that journalism is *not enough* that leads to problematic praxis because our episteme is subsumed by the powerful logic of a neighboring field. In particular, journalism of attachment has been linked to humanitarian military interventions by western powers, with journalists consciously campaigning for them (O'Neill 2012).

The issue I see in such campaigning[6] by foreign correspondents is that it does not foster the concept of a global public sphere, but rather absolves audiences of the responsibility to make up their own mind. As noted in the introduction, "A global public discourse does *not* arise out of a consensus on decisions, but rather out of *disagreement* over the *consequences* of decisions" (Beck 2009: 59). Journalism of attachment does not provide this space.

Instead, as with the transactional frame, a ready-made solution is offered. This sidelines the development of the demotic global public, damaging the font from which we derive the very reason for our existence, and blurs the boundary between foreign correspondents and the policy world, a boundary that is already a dangerously porous place.

Finally, "journalism of attachment," by suggesting we need do "more than journalism," does unnecessary harm to the concept of bearing witness—unnecessary, because objectively bearing witness is already "hugely passionate, requiring emotional engagement and human imagination" (Wahl-Jorgenson and Pantti 2013: 197). Bearing witness is also an act that involves "responsibility and commitment toward victims of war and other forms of suffering" (Tumber 2013: 61). The tragedy in adopting the so-called more-than-journalism approach is that it seems to forget that most people involved in conflict, combatants and noncombatants on both (all) sides, are victims of it. Wright raises the point that the journalism of attachment can dehumanize those positioned as oppressors and "also points towards another problem. That is the inherent contradiction of the discourse of human rights and the dehumanization of aggressors" (Wright 2018: 148).

Bearing witness needs to be reaffirmed and celebrated as more than enough for foreign correspondents operating in a global public sphere. Commitment to that global public sphere, a commitment to both our interviewees and our multiple audiences, means that showing, not politicking, is the more ethical practice for foreign correspondence. This showing gives the audience greater scope to *engage* with the circumstances on which we are reporting and connect with our shared humanity.

Cultivating Complexity

> When someone's speech and action are not recognised, their speech and actions are made meaningless, and they are treated and judged, not according to who he or she is (through their words and deeds), but according to their membership of a particular category of person. This refusal to recognise someone's individuality, their unique distinctiveness is a refusal to recognise a fundamental aspect of their humanity and is profoundly dehumanising.
>
> —Fiske 2016: 31

In completely separate studies, when sociologist Lucy Fiske and anthropologist Michel Agier asked refugees in detention centers and refugees in camps about protests they had taken part in, they heard the same answer: refugees engaged in protests to remember that they were human—for the human condition is political. Fiske's interviewees stated that otherwise they were mere bodies or even objects. As eloquently put by one of her interviewees: "If I didn't do those things, nothing different between me and this table" (2016: 30). However, both academics also found that these political acts needed recognition that was often denied because they were misunderstood. Moreover, Agier raises the issue that this acknowledgment of refugees as political subjects, instead of objects-victims, is problematic because "If the subject is stateless, what they express then demands a space of political legitimacy beyond or outside the sphere of the nation-state that is globally recognized as the space of political right" (Agier 2016: 152). By moving their words and actions into an imagined global public sphere, foreign correspondence can provide this space of legitimacy. But it is not enough to simply report the words. The words are still meaningless unless we also try to provide enough context to make those words and actions understood.

Complete professional identification with the knowledge creation potential of bearing witness demands of us preparing fertile ground in which to sow reception of these borderland realities—and the human potential residing therein. That the quotations provided by our interlocutors can be understood, and not just heard, requires an understanding among audiences first that "there is no common world without alterity" (Agier 2016: 156) and then that the world is knotted and intricate. Stories of complexity (layering not simplifying) and connectedness (explaining the glocal nature of events, issues, and humanity) are the order of the day. We must also consciously cultivate the sense of space-time in which we are operating—both for ourselves and to convey to our audience. The borderland must be understood, through journalism, as a shared global present that has been crafted by history, rather than escaping from an incomprehensible lacuna. As suggested by Hannerz, "It probably matters less whether [a foreign correspondent] is in a place for a brief or an extended period, only once or many times. What makes the difference may be simply a cultivated sensitivity toward the passage of time: the medium-term history of the present as a state of mind" (2004: 230). Robert Fisk makes a similar point when he advises younger colleagues: "Don't just go there and report it as if it's a crime story. Take a history book" (quoted in McLaughlin 2002: 15).

Problematic parachutists cannot be exempt. They must be held to the highest standards, and, in a more positive view of their praxis, their worldwide itinerary does provide them with unique insight—the insight of comparison. Likewise, investment must be made in understanding the actually existing conflict- and non-conflict-based economies of borderlands. Understanding, and documenting, the dynamics of our *glocal* economy and its effects on people is not easy, but it is exactly the role foreign correspondents should be playing. It is this sort of journalism that can make the metalinks between places in the globe. Networks of local journalists and foreign correspondents, working on global stories, are perfectly placed to make this complex reality understood.

While documenting metalinks of the economy, it is also crucial to keep in mind the impact of powerful global discourses. For example, since the September 11 terrorist attacks in 2001, development has been linked to "enlightened self-interest" in a problematic and powerful security discourse (Duffield 2007: 3; see also Power 2006: 25), and in a discursive process of displacement, terrorism has come to be linked to refugee movements, and especially Muslim refugee movements (Ghani and Fiske 2020; Giotis 2021).

So is it any wonder that when I discovered that the man holding people hostage at gunpoint in a cafe in the center of Sydney, the perpetrator of the "Sydney Siege," was an Iranian political exile, my reaction was "Oh f*** he's a refugee"? I feared the strengthening of the "refugee = terrorist" equation. Of course things didn't go to script; a single face-to-face interaction that was turned into narrative and broadcast via a tweet created a totally different discourse around the event. When a woman in Sydney saw another woman taking off her hijab, for fear of being subject to reprisal attacks while riding home on public transport, she offered to ride with that woman to her destination. People in Sydney repeated the offer via the #illridewithyou and this response of love against the discourse of hate and fear was a mobilization that went transnational. No physical rides to work were shared the next day in other parts of the world, but the message was shared, and "#illridewithyou" became a cellular organization—lone wolves of hope—deployed in response to that other cellular organization—ISIS—and its lone wolves of despair— the clash of transnational cellular organizations of terror and utopianism foretold in Appadurai (2006: 137).

What does all this have to do with the epistemological basis of foreign correspondence? Simply that we *co-create* every moment of knowledge with our intended audience. Entman points out that "news organizations shape their reports to elicit favorable reactions from readers and viewers" (1991: 7). Consider too, Lederman's conclusion after comparing coverage of the first Iraq war and the first Intifada. He states that "more than anything else" the nature of the coverage is "the product of whom the decision maker or the journalist chooses as his or her target audience. Once that choice is made, each participant in the event becomes a captive of his or her own perception of what the chosen audience wants, expects to hear and see, or can be made to believe" (Lederman 1992: 328).

Luyendijk draws a similar conclusion when comparing European and American journalists' coverage of the toppling of Saddam Hussein's statue with the coverage of the same event by Al Jazeera (2009: 7). Yet modern media ethicists have sought to complicate and challenge these practices by pointing out news media are now "global in content, reach and impact" (Ward 2013: 8) and that parochially minded media "can wreak havoc in a tightly linked global world" (Ward 2013: 2).

For foreign correspondents in particular, the reality of globally accessible media means that greater emphasis is placed on their responsibility to local audiences as part of the co-creation process—our representations have real effects on those we are reporting on. Take, for example, the Anglo-American media's culpability in the creation of Afropessimism. Surely it is no coincidence that when Zambians living in that country's copper belt discussed their problems, and in particular their sense of abjection and debasement, they changed register to refer to themselves not as Zambians but as Africans, an identification that "evoked all the images associated with Africa in contemporary international media discourse—pictures of poverty, starvation, and war; refugees, chaos and charity" (Ferguson in Hannerz 2004: 132).

Negative effects on local audiences are not a foregone conclusion. Wahl-Jorgensen and Pantti note that "global problems require global solutions, and journalistic practices play a crucial role in both enabling and limiting particular outcomes" (2013: 192). The Frame Reflection Interview is one practice that can play a role here. When introduced to a working correspondent's practice, the correspondent reported feeling a greater sense of responsibility both to multicultural local audiences in Australia and to the local audience in Indonesia where they were reporting from. Among other points, this sense of responsibility manifested in a determination to "come to grips with the complexity [of issues] and strive to report it" (Giotis and Hall 2021: 12).

Narrating complex stories can be achieved. This task is helped if there is affirmation of the value of frame and discourse diversity, within the journalism profession more broadly, and in the professional identity of foreign correspondents specifically. Moreover, this professional identification requires a commitment to understanding the complex web of co-creation we operate in. Consider a foreign correspondent (FC) in a refugee camp in Goma. Their web of relations includes FC-place (camp plus Goma plus DRC); FC-(I)NGOs; FC-interviewees; FC-home office; FC-society of home audience; home society-ideas of that place; home society-ideas of itself. To take just the final strand in that web, imagine what stories we will tell of Mugunga camp, and in particular the way it is intertwined with conflict, despair, violence, and international people movements, if we look to our audience and see a society besieged, or, as happened post-Sydney Siege, we look and see a society actively and confidently discussing the redeployment of Australian multiculturalist identity as an antiterror tactic ("Gruen XL" 2015).

The likely scenarios go thus: If the correspondent is aware the discourse in the Australian polity is one of fear and a drawbridge mentality, then a foreign correspondent reporting for an Australian audience is likely to see Australia as disconnected from the world. If such correspondents also have no sense of themselves as having a professional episteme that is meant to challenge that sense of disconnect, that is, they do not possess professional injunction to craft stories of complex interconnectedness as part of their foreign correspondent identity, then the likely story to emerge from Goma is a simple one of compassion around Mugunga camp, perhaps in the transactional frame. By contrast, if the correspondent is conscious that there is also a discourse in Australian society celebrating multiculturalism as an achievement to be shared with the world, and the foreign correspondent also has a professional identity imbued with the principles of a global polity, then perhaps the story to emerge from Goma would be a complex story of conflict, perhaps told through communities reconciling after interethnic violence. In my own reporting of the Sauti group, if I had a more developed awareness at that stage, I could have done a story like that because one of the organizations, Fupros, worked exactly on this complex issue.

However, and this is a big *however*, the problem with everything I have said so far regarding complexity as part of the professional praxis of foreign correspondence is that accessing both information and various discourses is the *easy bit*. The discourse may not be strong and it may not resonate with our audiences, but it is still easily accessed. The really difficult part is the micro bit—building up the discourse so that it does resonate. This has to be done one piece of medium at a time, by first researching the stories and then finding appropriate frames and narratives for individual pieces of foreign correspondence within the limitations of the genre of journalism.

Take, for example, this paragraph discussing the Lord's Resistance Army in the 1990s:

> Their rebellion is political inasmuch as it is a result of the world political and economical system in place. It is, for those engaged, non-political inasmuch as they no longer believe in politics. For the majority of the rank and file, it is a survival strategy, a way to obtain things which are out of reach by all normal means: consummatory rewards as ideological drive. For those in charge, the brokers, it is a tool to acquire access to power, status and some wealth: rebellion as a career. One of the bitterest observations is that, although they turned against the system, they are still used by that system. Whether it is a state, a political faction or a private company which uses them for its own purposes, the rebel movements are bought and sold on the market. And if the price they receive for their final destruction is low, it is a reflection of the logic of the world market. (Doom and Vlassenroot 1999: 36)

It would be quite a feat to take that paragraph and turn it into engaging journalism while maintaining all that complex interplay. It is much easier to write an adventure-frame story about meeting the warlord Kony, which, as discussed in chapter 1, too often happened. And yet a commitment, as part of the foreign correspondent episteme, to conveying this complexity means we must find a way. And perhaps, if it is too confused for one story, as "The Congo Connection" was, that simply means that we must create more stories and convey this interconnectedness reaffirming the importance of foreign correspondence as a key form of journalism.

This type of foreign correspondence will generally require more than a shoestring budget. I haven't at any point forgotten the economic element. I simply posit that there is nothing to be gained from viewing the economic crisis in journalism as an insurmountable barrier. All cultural industries face the increasing and constant challenge of the monetization of their work. The economic crisis means that the media world is in transition, and this is exactly the right time to consider issues of identity.

Professional identity is a driver of decisions in any professional group—indeed it is the foreign correspondent hero identity that maintains the expense of parachuting and datelines. I follow Reese in arguing that an authentic professionalism would also act as an awesome counterweight to the cost of global newsgathering. So while I acknowledge the economic constraints, I do not think they are the only consideration. In fact, a more significant obstacle than the financial one may be our adherence to the norms of the genre.

Speaking specifically of the genre of war reporting Boyd-Barrett has noted:

> The genre plays into the hands of power, and this is nowhere more apparent than the media's failure to identify the metanarratives or grand strategies that explain the links between different wars over extended periods of time . . . its generic character has been exploited by state and other propagandists in ways that cripple the capacity of media consumers to make useful sense of the world. (2004: 25)

Boyd-Barrett argues the DRC's conflict has been ignored by western media, despite significant western involvement and interests (a list of which he provides), because of the involvement of proxy forces, which do not slot easily into the typical genre of war reporting (2004: 27). If this is correct, we are failing to convey complexity and also failing to create the global public sphere because we are not expanding the knowledge production capabilities of the genre of war reporting to suit new realities.

If foreign correspondents are indeed blinkered by the norms of the genre, then the correspondents themselves may be helping to keep the blinkers on by arguing their reporting faults stem from audience limitations. Carruthers explains that "in journalists' telling [it is] Western audiences [that are] conceived as intolerant of complexity, especially in places of which they know, and care, rather little" (2004: 165–66). I would argue, with de Botton (2014), that we have no business blaming the audience, and just as we need to cultivate the capability to care through a focus on everydayness instead of exception, so too we need to cultivate the capability to want to delve into complex phenomena by building complexity into our narratives as a matter of course. In fact, I would argue that these two cultivations—caring and complex understanding—are symbiotic, and that both are intertwined with foreign correspondents' unique value as agents of normative cosmopolitanism.

Professional Cosmopolitans

An ordinary cosmopolitan condition is being formed . . . in a global and hybrid world . . . *on* the border, that is, in everything that makes

> for the border . . . landscapes in which encounters and experiences
> bring into relation a here and an elsewhere, a same and an other, a
> "local" fact and a "global" context.
>
> —Agier 2016: 8

Frantz Fanon's experience and analysis of colonialism were productive of
more than knowledge of the world as it was—it also produced a postcolonial cosmopolitanism infused with humanism. His writing was heterotopic
in that he was able to describe the contradictory clash between the discourse
of French Enlightenment ideals and the realities of racial inequalities in the
empire. It highlighted the "contradictory cultures" of colonialism and the
way in which the very fissures of the system "carried its own distinct logic of
cosmopolitanization" (Go 2013: 211). In borderlands today similar contradictory cultures exist; the human rights discourse based on equality is belied
by the racial hierarchies of biosecurity. Without a commitment to normative
cosmopolitanism it is impossible to understand, and communicate, this contradiction and it is impossible to find its productive seams—the complex,
glocal stories of everyday resistance so desperately needed.

If we accept the preceding argument, then what will be the journalistic
tools of choice? Important work has already taken place in this area by
looking at current crisis reportage. Cottle argues that the possibility for a
global public sphere developing through news of global crises is stymied,
as "actions and reactions are often reported in and through national news
prisms and frames of reference" (2009a: 509). This sort of nationally based reporting also stymies the potential for normative cosmopolitanism. Examples
of this stymying include "reporting on distant disasters and humanitarian
emergencies [where] national news media continue to seek out and populate
stories with their own 'nationals'—whether embodied as victims, survivors,
heroes, or concerned celebrities" (2009a: 509). Cottle's research suggests
avoiding frames built around the "national," and my research would suggest
avoiding frames that activate the Africanist discourse.

The question of which frames to use is perhaps a little harder to answer,
but work has been done. Berglez suggests that a "global outlook" is fostered
when global relationships are central to the reportage, when the inescapable
nature of global issues is emphasized through concreteness, that is, the ability
of journalists to demonstrate in their stories concrete examples of globalization in people's everyday lives (Berglez 2013: 20–49). Cottle and Rai suggest
that different frames within the common communicative architecture of TV

can be analyzed for their "impact on the mediatization of major global issues, including ecology, the 'war on terror,' population flows, indigenous and
minority representation, and human rights" (Cottle and Rai 2006: 185), but
Cottle also notes "there is no single form of reporting or universal cosmopolitan frame that can be prescribed in advance" (Cottle 2013: 244). After
all, each story has to take into account the actual realties of that particular
situation. Nevertheless, I would like to suggest the more we focus on and
build up examples of potential frames, the better. This research, via the literature I have examined, and via the process of frame reflection both through
my FRIs, and through the writing of the "Furaha" piece, suggests that our
"shared humanity" may be a useful frame to add to our common reper-
toire, especially if it is properly differentiated from the humanitarian frame.
This also aligns with research conducted by Nothias (2020), who found that
foreign correspondents who were interested in challenging problematic
representations of Africa "championed the use of human-interest angles in
their reporting to promote a sense of common humanity" (2020: 257).

Our thinking around frames may also need to be complemented with
thinking around format, for the format within foreign correspondence that
seems to allow the most scope for both normative cosmopolitanism and
complexity is the feature. The feature article (or feature report on TV/radio)
seems the obvious choice for stories of complexity and interconnectedness
because it has "the space and relative freedom—compared with 'straight reporting'—to demystify complexities most readers find forbidding" (Conley
and Lamble 2006: 315). The format also allows the use of "comparison and
contrast, surprise, analogy, metaphor and sensory appeal" (Conley and
Lamble 2006: 316) and these are all devices that allow the audience greater
engagement with the story, remembering that Aristotle called metaphor "the
most pleasant form of learning" because it actively involves the audience
(Aristotle and Kennedy 2007: 218–19).

Feature articles are noted as a strategy for shifting stereotypes by
correspondents themselves (Nothias 2020: 258) and also endorsed by
Hannerz in his discussion of foreign correspondents as agents of normative
cosmopolitanism. He notes that features allow for a "range of sentiments—
surprise, empathy, amusement, irritation—much as life generally offers" and
"can contribute to a cosmopolitan stance, a cosmopolitanism of the thicker,
denser, richer kind" (Hannerz 2004: 33). Hannerz gives examples of features
that, instead of othering, give the sense that "these are our contemporaries
out there" (2004: 34). He argues features center individual people with all

their nuances, and this allows us to see that "culture tends not to be a long-durable consensus but a shifting although sometimes distracted debate" (2004: 231). Hannerz then contrasts this sort of nuanced understanding of culture to a thesis like Huntington's *Clash of Civilizations*. I find this contrast interesting and I note that Huntington's thesis sees the world as *divided*, so the corollary would suggest that a more nuanced understanding of global cultures and people would help create a more *unified* sense of the world—a global public sphere.

A more conscious professional commitment to the feature format as part of the professional craft of foreign correspondence means examining past and current practices. Wright (2018: 248) notes that INGO-produced material often finds a home in off-agenda news features, so current practices may be privileging INGO viewpoints. Moreover, the feature format is probably overused for "happy" stories and infotainment, perhaps as an attempt to balance the negativity of hard news. Hannerz notes that a spate of feature articles from abroad in the 1990s dealing with the spread of American culture left Americans profoundly unprepared to answer the question post–September 11, "Why do they hate us?" as anti-Americanism did not receive the same sort of coverage (2004: 200–201). Writing a feature doesn't mean searching for happy stories, but it does mean remembering to bear witness to our shared humanity and there are techniques for doing this. The framing of my story about Furaha around teenage practices, worrying about clothes, dramatic remonstrating with parent's decisions (Kahumbu's children), spending time on hair—these were conscious choices in the same way that award-winning cameraman Glenn Middleton says: "I try to capture sound and images that make viewers sit up and say, 'Hey, I can relate to that. My kid does that'" (Leith 2004: 265).

Furthermore, it is not a case of relying on the feature format—and storing up all our representations of complexity, interconnectedness, and shared humanity for that format. This would be a mistake. Supplements and documentaries are presented in a way that conveys the impression that they provide "optional" "background information" and that you can still understand the world if you follow "just the news" (Luyendijk 2009: 99). Yet, as Luyendijk points out, feature articles are not really optional; they can help explain the world in a way that a digest of news headlines (often stereotypical) can't.

I am arguing that we should not wait for the opportunity of a feature format to insert a human, mundane moment. By not waiting for the opportunity of the feature to portray everyday moments and complexity, we would in fact create a virtuous circle that would incentivize the production of more features. Seeing normative cosmopolitanism and complexity as part of the foreign correspondents' knowledge creation process would mean we were more likely to include instances of the everyday and/or complexity in *all* our reports. This would build up audience capacities and it would mean that reading complex, long-format stories would be less daunting. This would create more demand for features as part of the craft of the foreign correspondent—the repetition of this practice would then help embed these types of normative cosmopolitanism stories as part of the foreign correspondent identity—and now we are back to the beginning. The circle repeats and strengthens.

Before leaving this discussion of normative cosmopolitanism, I wish to clarify that the type of normative cosmopolitanism I am referring to throughout this section starts with Beck's definition of "*recognition* of others as equal *and* different" (2009: 57) and is *not* the same as compassion. This distinction is not always made in literature on foreign correspondence. The normative cosmopolitanism I am describing aligns with the definition drawn on by critics of Aid/Peaceland who argue not for compassion but rather for a "sense of universal humanity, [and] reflexive distance from one's own culture and relating to traditions other than one's own" (Anderson in Harper 2011: 124). Thus, it meets the criteria of Ward, who argues that "we need a cosmopolitan media that reports issues in a way that reflects this global plurality of views and helps groups understand each other better" (2013: 2). In my definition of normative cosmopolitanism for foreign correspondence, I also return to the Greek roots of the word: *cosmo* (world) and *polity*, clearly pointing us to the fostering of a global public sphere.

This distinction is necessary, as in the literature on foreign correspondence the terms "cosmopolitanism" and "compassion" tend to be conflated, a problematic practice given that the two types of framings lead to different discourse outcomes. Normative cosmopolitanism corresponds to a discourse around a global public sphere, whereas compassion is too often tied to a discourse around the white man's burden. An example of this conflation is when Wahl-Jorgenson and Pantti discuss the "role of disaster reporting in engendering *compassion and, relatedly, a cosmopolitan sensibility*" (2013: 193; my italics). The authors then enter into a discussion of journalists contributing to "humanitarian action" (2013: 195) and the discussion then slips further down the white man's burden slope to the shallow transactional frame via this quotation from a BBC editor/reporter, who opines of the audience: "While you're complaining because your McDonald's is a bit too hot, there are people dying that could be saved by the 2 pounds you spent on that coffee" (2013: 195). This slippage, from thinking about the development of a global public sphere to talking about the transactional frame, once again devalues the episteme. Another example is when the UK's Channel 4 News correspondent Jonathan Rugman asks, "How are you going to make your viewers care, which they must if they are to donate money, which of course *makes our jobs that much more worthwhile?*" (Sambrook 2010: 30; my italics). This is not to undervalue empathy as a force in the world; of course this is a capacity foreign correspondents should foster, and this takes skill to do, but in no other reporting situation would journalists say their job is fundraising. Such a comment points to the urgency with which the purpose of foreign correspondence must be reimagined.

If the type of cosmopolitan identity I am espousing has no space for the problematic practice of the white man's burden, it also has no space for the current disrespect for language and specificity. The disrespect inherent in the lack of language skills is discussed by both Mbembe and Lederman with Mbembe writing:

With some analysts, only reading French, others only English, and few speaking local languages—the literature lapses into repetition and pla-giarism . . . distinguishing between causes and effects, asking the subjec-tive meaning of actions, determining the genesis of practices and their interconnections: all this is abandoned for instant judgment. (2001: 9)

Mbembe points out the distorted understanding of the world resulting from the lack of language skills, and Lederman goes further, highlighting how this works in the practice of crafting stories. Lederman argues that it results in

entrapment in particular storylines, a distortion of events being covered, a self-reinforcing overemphasis on the dramatic and non-verbal aspect of events, and weak, one-sided analysis based on the ruminations and often baseless or self-centred prophecies of the elites rather than on the thoughts and concerns of real, powerful, but English-mute constituencies. (1992: 123)

In contrast consider Burchett, who was not only known for being the first correspondent to report from Hiroshima, he was known for his "early self-training in foreign languages [which] put him a mark ahead of other journalists" (Shineberg 1986: 139) . This training enabled him to learn the local language for his acclaimed reporting on New Caledonia.

When I entered the journalism profession, it was already suffering from rapidly falling revenues, and staff training was less of a priority, but my organization still offered shorthand training and we were encouraged to take it up with the motto "Better quotes make better stories and better notes make better quotes." Why does that value go out the window when it comes to foreign correspondence? We should not ignore the fact that people in countries visited by correspondents may speak English but "think in their mother tongue. The words, metaphors or myths that most accurately express their thoughts are best fashioned in their mother tongue" (Lederman 1992: 122). A commitment to normative cosmopolitanism with its "different but equal" mentality would at least help to make the argument for language investment. But what I really want to point out is that this kind of approach, an approach that respects alterity, finds depth of expression in the philosophy of Emmanuel Levinas.

A Professional Philosophy of Praxis

> The Other who expresses himself is entrusted to me.
> —Levinas in Derrida 1999: 7

Everything I have written so far in this chapter speaks to an ambitious agenda and, accordingly, we will stumble. Our individual practice will fall short again and again and we will find ourselves in hotel rooms, raw, resenting luxuries, our seemingly apathetic audience, our editors, and the media machine—once again questioning "the thin dividing line between reporting and exploiting, about using desperate people as raw material" (Buerk 2004: 292). When this happens, as it is bound to do, what will we then fall back on? When the horror of the world overwhelms and our journalism does not get the audience response we want, which will happen again and again, how will we resist the temptation to once again seek "more than journalism"?

I return to Bourdieu's point that "the social world doesn't work in terms of consciousness; it works in terms of practices" (Bourdieu and Eagleton 1992: 113). What we require most of all is a strong philosophical underpinning for our praxis that reflects the intrinsic value of bearing witness, of being there, and of respecting alterity.

Emmanuel Levinas is a philosopher who dealt with foundational ethics, "ethics of ethics" (Derrida in Bergo 2015: online), meaning, in the context of his phenomenological philosophy, "the exploration of conditions of possibility of any interest in good actions or lives" (Bergo 2015: online). Levinas was a Jewish Lithuanian, a resistance fighter, and then a prisoner of war in World War II while his family died in Auschwitz. Levinas became a key intellectual figure in France, described as someone "without whom philosophy would not have been what it is" (Marion in Critchley 2002: 2). Levinas's philosophy deals with the "lived experience" of the world, and especially the "rise and repetition of the face-to-face encounter" with the other (Bergo 2015: online). A more perfect philosophy for the praxis of bearing witness can hardly be imagined.

I am not the first scholar concerned with journalism to be drawn to Levinas. His philosophy inspired Roger Silverstone's (2013) development of the concept of the proper distance of reporting, where the other is neither so distant, that we become strangers, nor so close that difference is ignored. Levinas was also an inspiration to Judith Butler's work *Frames of War* (2016) and how the media constitutes precarious lives as faceless. Butler was in turn an inspiration for the decolonizing journalism scholarship of Lindsay Palmer (Palmer 2018, 2019), who places her work as part of the ethical injunction posed by Butler, that critical scholars must "frame the frame" (Palmer 2018: 162). However, my first introduction to Levinas was in research about Australia's treatment of asylum seekers and its relation to the following point: "Levinas's philosophical treatise insists that each individual being is unique and responsibility is dialogical relating in a face-to-face encounter with the Other. Listening and relating *with* must precede thinking *about*" (Wainer 2010: 5). In Levinas's exhortation to relate *with*, before thinking *about*, I found coherence with concerns about the distancing effect of victim framing. This exhortation against thematization of the other is an important part of Levinas's philosophy: "To think the infinite, the transcendent, the Stranger, is hence not to think an object" (Levinas 1979 [1969]: 49). Levinas sees the other as an interlocutor and believes that the encounter is capable of creating a just, ethical, language of plurality that overcomes rhetoric. In contrast, "Thematization and conceptualization, which moreover are inseparable, are not peace with the other but suppression or possession of the other" (1979 [1969]: 46). What *Totality and Infinity* (Levinas 1979 [1969]) teaches us is that only by respecting people's alterity, again, and again, and again—infinitely, are we safe from thematization and the "totality of history"—genocide.

But we have failed in our ethical responsibility if we start thinking about these issues at the writing-up stage. We must begin with our interactions—especially with our interviewees—we must view each encounter as being with an interlocutor whose alterity we respect absolutely. As with bearing witness, Levinas gives primacy to physical proximity, positing ethical relations, (wo)man to (wo)man, as the structure on which all other structures rest. Meaning, he argues, is "said and taught by presence" (1979 [1969]: 66). Further, primacy is given specifically to the face: "The true essence of man [*sic*] is presented in his face, in which he is infinitely other" (1979 [1969]: 290). Levinas goes on to say that this presence of the face "arrests and paralyzes my violence by his call" (1979 [1969]: 291), which we can also take to mean the violence of thematization.

This philosophy of the act of interacting with others presents a complete contrast to the following description of journalists in the DRC:

> According to an observer who accompanied the UN representative Margot Wallstrom in visiting the victims of the mass rapes in Walikale in 2010, the convoyed group of journalists rushed out from the helicopter upon landing in a frenzied search for victims to interview, and later compared and argued over who had been able to document the worst case. (Baaz and Stern 2013: 92)

The journalists involved were not evil, but they had obviously given in to the temptation of thematization, and specifically the "has anyone here been raped and speak English" thematization. This take on the situation, driven by the limits of the genre, has a long history of repetition in correspondence,[7] and even deeply reflexive practitioners admit they will "think [it] at times" (Lorch in Leith 2004: 233). Levinas's philosophy helps us understand the deep ethics involved in the praxis of *not* rushing out of the helicopter looking for themed others, and in that understanding correspondents could find support to fight that temptation when it arises.

On the subject of normative cosmopolitanism, Levinas is also useful. Paralleling Beck's discussion of normative cosmopolitanism as *one* possible response to the realization of an interconnected world of risk, Levinas posits *respect* for the absolute alterity of the other as *one* possible response to the challenge of the face-to-face encounter. Levinas highlights that the encounter with the other, which is necessary for cosmopolitanism to be achieved, is intrinsically challenging because the very existence of the other involves the subsuming of the "I." But it is only absolute and infinite respect for others in all their alterity that keeps us on a firm ethical footing. Levinas gives us an argument as to why we should strive for this normative understanding of cosmopolitanism:

> Acute experience of the human in the twentieth century . . . [means we cannot deny the fact that] . . . hunger and fear can prevail over every human resistance and every freedom! . . . Freedom consists in knowing that freedom is in peril. But to know or to be conscious is to have time to avoid and forestall the instant of inhumanity. It is the perpetual postponing of the hour of treason—infinitesimal difference between man and non-man— that implies the disinterestedness of goodness, the desire of the absolutely other. (1979 [1969]: 35)

Levinas draws a distinction between desire and need. Levinas states: "Desire is an aspiration that the Desirable animates; it originates from its 'object'; it is revelation—whereas need is a void of the Soul; it proceeds from the subject" (1979 [1969]: 62). In this, Levinas suggests that of the two, desire is the more ethical stance. This becomes relevant for us when considering the difference between the enforced enlightenment of the realization that we live in an interconnected world of risk, and thus *need* a cosmopolitan understanding, versus the *desire* for normative cosmopolitanism. The philosophy underpinning our craft must *desire* normative cosmopolitanism. It is a failed episteme, and simply more media management, if a turn to normative cosmopolitanism is motivated not by ethically infused search for knowledge but instead by a survival strategy justifying the role of correspondents. To be clear, it is both, but the ethical desire must take priority. To work with need rather than desire puts our ability to see the absolute alterity of others in jeopardy—the animation of their alterity must proceed from them.

Finally, Levinas's philosophy gives us a way to work through the survivor guilt of bearing witness—that is, the burden of seeing unfairness and tragedy, and knowing that you have the luck of being able to walk away. That same emotion is felt passing any person, in any city street, with outstretched arm or placard, except this action of passing by, and the emotion of survivor guilt, is multiplied a million times over in any necropolis. As foreign correspondent Sally Sara has noted in relating her struggle with post-traumatic stress disorder (PTSD):

One of the big misconceptions is that PTSD is all about the blood and bullets and the bombs and the fear. But, for many people it's more complicated than that. Mental health experts use a term called moral injury. That means that what you saw was not just physically confronting, but it was wrong, morally wrong. (Sara 2014: online)

In acknowledging that feelings of moral injury and survivor guilt are a *natural* part of the praxis of foreign correspondence, a kind of parallel can be drawn to teachers tackling the topic of disadvantage, where they often witness the "rage, embarrassment, and genuine passion through which students express their guilt and innocence in becoming aware of the suffering of others" (Todd 2003: 104). Traditionally, in the classroom, this guilt has been viewed as unhelpful, but Todd seeks to draw a line under the unhelpful concept of liberal guilt, and instead, leveraging Levinas, she reframes guilt as moral orientation of awareness and susceptibility (2003: 97–107). Todd argues that "it is the vulnerability and susceptibility *to* the Other that Levinas views as significant for being responsible *for* the Other" (2003: 100). The same sort of reframing could be applied to the concept of bearing witness.

Consider the common experience of barroom debriefings. These debriefings often occur after those involved have been in physical danger or have witnessed moral affronts. Telling, out loud, the story of physical danger and survival can be helpful in overcoming the emotional toll taken when going against human instinct and running toward danger. At the same time, these debriefings provide "vocabularies of motives" and "acceptance frames" just like those Wacquant documents in the pugilist community (1995: 83–85). Wacquant finds boxing gym conversations help reinforce the doxa of a dangerous activity; likewise, bars provide a space for foreign correspondents to normalize dangerous and unusual work practices. More unhelpfully, when these stories are retold outside the foreign correspondent community, shared in bars far from borderlands with the wider journalism community, there is also an element of unconsciously activating the white hero stereotype.

So much for the days that involve physical danger; what about the days that involve moral affront? Here, barroom tales seem less positive, for, as in the example of "comparing and arguing" over "who had been able to document the worst case" (Baaz and Stern 2013: 92), they can only cement the moral confrontation by literally adding insult to injury. I suggest these problematic barroom tales are driven by the view, much like in the classroom, that guilt is an unhelpful and disabling emotion, and so foreign correspondents try to drive it away with cynicism. In the process of course, it is simply driven deeper within—stored up to explode another day. Could understanding the *necessary* susceptibility involved in our responsibility to the other change bar stories to discussions of guilt and vulnerability, not only as a balm for the soul, but in beautiful counteraction to the hero trope?

To reiterate, this does not mean those engaged in foreign correspondence will now go forth without stumbling, or that all future works will overcome the problems of the past. Even if foreign correspondents think they have been as ethical as possible in their praxis, the work may have a life of its own. Indeed, Levinas argues we cannot be understood based on our works: "From the work I am only deduced, and am already ill-understood, betrayed rather than expressed" (Levinas 1979 [1969]: 176). Yet we can understand our epistemological praxis as being rooted in the deep philosophical ethics of the face-to-face encounter. This is the strong foundation from which, as humans first and *then* as professionals, we will be able to try again and again.

I am advocating a professionalism that cultivates heterotopic knowledge and imagination, takes up the responsibility of creating knowledge of borderlands through bearing witness, crafts stories of glocal complexity and interconnectedness, adheres to principles of normative cosmopolitanism, fosters a glocally conscious global public sphere, and is underpinned by a philosophically informed ethical praxis. Surely this counts as an "authentic professionalism," as sought by Reese at the beginning of this chapter, and surely these are the storytellers we would wish to read and be.

Conclusion

Flaming the Fanonian Spark

Perhaps the achievement of paradise was premature, a little hasty
if no one could take the time to understand other languages, other
views, other narratives period. Had they, the heaven they imagined
may have been found at their feet.
>   —Toni Morrison on the Tower of Babel (Nobelstiftelsen 2006:
>   187–88)

Among Fanon's many intellectual gifts to the world was a demonstration that
the structures of today need not be the structures of tomorrow. As discussed
by Go (2013), even among the violence of colonialism, in fact through that
violence, Fanon developed his thesis for a strengthened humanism, in part
using the tools of the colonizers against them. Human action takes place
within social structures. At times agency within these structures resembles
a decision between a poor set of options you wish you did not have to choose
between—in that sense there is an element of coercion—but there is still
agency. Most interviewees in borderlands are there because of constrained
choices. That is not the case for those journalists engaged in foreign correspondence. I *freely* chose to go the Democratic Republic of Congo. This free
choice does not mean I was free of structures. The experience was structured
by the social geography of the borderland, by my professional concerns, and
by myriad other social structures I carry inside me. For me the danger was
not the limitation of my agency but that I would misrecognize structure as
my own free will. I might unreflexively use those structures to further my
professional ends without understanding the effects of my correspondence
on my interviewees.

The ideas in this book have been broad ranging because the work of decolonization is environmental, by which I mean it encompasses all aspects

Borderland. Chrisanthi Giotis, Oxford University Press. © Oxford University Press 2022.
DOI: 10.1093/oso/9780197565797.003.0008

of our lives as professionals and human beings. The ideas in this book are also driven by the realities of borderlands. This is not the time for small targets. For decades we have been living in *Pax Imperii*, "a false pretense of peace that really presides over a state of constant war" (Hardt and Negri 2004: xiii). In 2020 Syria was the top "source country" for forcibly displaced people (UNHCR 2020). Its conflict has been predominantly framed, and understood, as a civil war. Left out of the frame is the impact of the global resources war on its doorstep. In 2011, before the war in Syria erupted, that geographically small country of less than 20 million hosted the *third largest refugee population in the world*: 755,400 people, displaced in the aftermath of the US-led attacks on Iraq. Left out of the frame are the millions of ordinary people around the world who fought against their governments to try to stop the Iraq war—the largest protests in Australia's history (Mercer et al. 2019 [2003]), the largest peace protests in world history, marked by the presence of people who had never before protested (Tarrow 2010). The knowledge of the demos was ignored. Left out of the frame too is the argument by Hardt and Negri that all wars should now be seen as part of a "global civil war" (2004: 4). The national space

> is no longer the effective unit of sovereignty . . . across the global terrain. . . . Each local war should not be viewed in isolation, then, but seen as part of a grand constellation, linked in various degrees both to other war zones and to areas not presently at war. (2004: 4)

What is learned from the case study of the DRC can be applied elsewhere. A significant realization of the lacuna effect is how missing information can be made to seem natural, and this has flow-on consequences. "Representing the Congo as a primitive, chaotic, 'heart of darkness' has made certain things happen in the political world" (Dunn 2003: 173), yet those representations, those frames, were not engaged in by journalists for cynical political purposes; they were and are used because they are outdated tools of the profession. The crucial decolonization process means understanding the power of framing. Furthermore, writing in frames, without considering power relations, unreflexively operationalizes hegemony.

The lacuna effect, put into political context, shows how knowledge is power and that this is most true in a relative sense. For example, the more people know about a country the less likely they are to support war or sanctions against that country, a fact that multinational oil company Exxon once sought to exploit (Harvey 2009: 101). Foreign correspondence must have a role to play in this power dynamic, a role that goes beyond pure information—for connection too is power—the power to imagine other ways of being in the world, to imagine ourselves out of the *Pax Imperii*.

The second key learning is that "the refugee is not a marginal figure, but the point where the limits of the modern political system become visible and an invitation to think to new politics" (Seth 2014: n.p.). Working in a refugee camp highlighted the role of Aid/Peacelands as part of the global rise of nonstate governance—a rise that foreign correspondence, given its crucial role in the global public sphere, should highlight. Adding a normative dimension of professional cosmopolitanism to a reinvigorated epistemology for foreign correspondency in a global public sphere, we come to important principles, for example, that storyline diversity is a crucial service foreign correspondents should provide in the global public sphere (Lederman 1992: 235). News stories, especially world news stories, function as gateways to the future (Hannerz 2004: 207). They need to be first drafts of histor*ies* so that more than one potential future can be envisaged.

Bourdieu and Harvey helped me understand that this new reality can *only* be built through practical elements like the conceiving of ourselves as heterotopic agents, the use of our "access all areas pass," of time spent in heterotopic space, and of stories aimed at fostering frame diversity, complexity, and interconnectedness. Both theorists emphasize practices. Similarly, both theorists in their different ways highlight the importance of "lifestyles"[1]—a particularly important issue for this research considering the role of lifestyle in the maintenance of Aid/Peacelands and biosecurity.

Practices are not only social habits, but also professional norms; the Frame Reflection Interviews, as a type of qualitative interviewing, fashion new norms for the purpose of exposing different knowledges and viewpoints. Picard suggests that the opposite of such diversity, the reinforcing of "dominant perceptions of issues, people and countries, especially those perceptions held by editors and social elites" leads to the media "consistently miss[ing] . . . major stories . . . until they emerged as full-blown calamities and disgrace" (2014: no pagination).The FRIs help reveal alternative histories, and presents, and act as a correction to the power dynamics that generally see subaltern visions for the future made invisible in the political public discourse.

Of course, it's not only journalists who have media-induced blinkers and stereotypes; audiences bring with them their own cultural baggage. Society's image of "over there" and its image of itself impact the co-creation process, highlighting the complexity of interactions between "there" and "here." The domestic/foreign split in audience perceptions and in the allocation of journalists covering different stories is responsible for poor coverage of refugee issues (Zelizer 2017: 75–76), but it needn't be this way. Chouliaraki argues that news journalism is a "*performative* practice, that is to say a practice that constitutes the communities it addresses at the moment that it claims to represent them" (2013: 268). This means that a global public sphere, a place for the political legitimacy of the almost 80 million stateless and displaced in the world today, is possible.

Possible is not the same as easy. It is here that the relational-based philosophy of Levinas becomes so valuable. *Totality and Infinity* is "regarded as one of the most significant philosophical texts produced in the second half of the twentieth century" (Davidson and Perpich 2012: 1). In a world where unending war threatens to become the new normal, Levinas's "radical heterogeneity" (1979 [1969]: 36) and the contrast he describes between totality and infinity help orient ideas about valuing alterity. I was prepared for the fact that his philosophy would provide intellectual depth to the synthesis of alterity. What surprised me was the contribution of Levinas's philosophy to a reinvigorated understanding of bearing witness. He provides an ethically, relationally charged understanding of going beyond thematization to draw beings out of the totalizing politics of history, "beings who always already speak" and whose speech is "always already a pedagogy" (Wyschogrod 2002: 191).

Levinas's philosophy, in application to professional practice, has not always been successful. One challenge encountered is in the nature of Levinas's philosophy, which, although described as "first ethics," is in fact not "normative but descriptive" (Bernasconi 2012: 253). This distinction has not been properly observed when Levinas's philosophy has been applied to practical professional contexts where people are seeking ethical guidelines for their actions. Perpich points out that Levinas's application to the nursing and psychotherapy professions has been taken as guaranteeing the ethical importance of compassion and caring for others (2012: 128). The ethical call of the face, which Levinas describes in *Totality and Infinity*, Perpich says, is turned into something "undeniable, like a fact" (2012: 141) rather than the moment of choice that Levinas describes. Perpich goes on to say: "Serious questions have to be raised about this kind of interpretation, which has become all too common in the literature attempting to move with Levinas into professional disciplines" (2012: 141).

Such interpretations, argues Perpich, make Levinas's account of ethics "too heavenly to be of earthly good" (2012: 134). And these interpretations are no doubt responsible for the critiques of Levinas in professional practice that see his philosophy as an "impossible burden" (Todd 2003: 16) and a "masochistic and self-annihilating gesture" (Todd 2003: 134).

The challenges encountered by other professions can serve as lessons. Bernasconi argues that Levinas himself has addressed the criticism of an impossible burden in the often-overlooked final section of *Totality and Infinity*, where Levinas introduces a discussion of time. Time is discussed as fecundity, multiplicity, and discontinuity, a structure of temporality that refuses totalization (Bernasconi 2012: 264). Says Bernasconi: "This is Levinas's answer ahead of time to those critics who subsequently characterized his philosophy of responsibility as excessively burdensome—from the perspective of infinite time, responsibility is not a burden" (2012: 266). Bernasconi further argues that "this does not diminish the responsibility, but clarifies its meaning, which is directed not to correcting the past but to opening up another future" (2012: 266). In a highly temporal profession like journalism, where we are always creating tomorrow's fish-and-chips wrappers, and where each new interview opens the opportunity for a new moment of respecting alterity, we need not be afraid of Levinas's "impossible burden."

Levinas provides, not rules, but an optics (Manderson 2012: 166), a worldview particularly relevant for our temporal moment. It has been argued that Levinas's philosophy is relativist in that it was forged in the moment of rupture created by genocide in Europe and, through its phenomenological descriptions, is "inviting us to an act of recognition" (MacIntyre in Perpich 2012: 150). Part of what we are being invited to recognize, says Perpich, "is the failure of traditional systems of moral thought that rely exclusively on rational principles" (2012: 150). In a world of necropolises the recognition of failure is once more urgent. As Manderson put it in discussing his engagement with Levinas: "Our theoretical resources seem often to get thinner the more thickly complex lives we lead" (2012: 169), the implication being that this shouldn't be the case. And surely this is especially true for foreign correspondence tasked with communicating the reality of complex, intertwined tragedies in a global public sphere.

Levinas's phenomenological philosophy should not be divorced from practices. I say this because it is quite striking to consider the extent to which this research has been helped by the mea culpa writings of former foreign correspondents in sub-Saharan Africa. Likewise, Nothias (2020) found in his interviews with 35 foreign correspondents in Kenya and South Africa a "postcolonial reflexivity," with many correspondents aware of postcolonial critiques and engaged in strategies to improve their reportage. We are not an unreflexive bunch. However, Nothias also found that, in an environment of structural constraints and precarious work conditions, some of these strategies ended up reinforcing stereotypes. Nothias argues: "Because postcolonial reflexivity appears to be a disposition that opens up a spectrum of possible practices, more work is needed in the future to understand the conditions and factors that encourage the adoption of different practices on this spectrum" (2020: 261).

This potential focus on identifying decolonizing practices is an exciting new theoretical coordinate in journalism studies and mirrors my motif, throughout this exploration, of focusing on praxis: that consciousness is not enough, that the social world works in terms of practices (Bourdieu and Eagleton 1992). It is only in the doing that knowing can be embedded in habitus. In the doing the ethical moment of decision, which Levinas speaks to, comes into being.

I know that this discussion of our professionalism is urgent when I look at images of reporters, regardless of any concerns about safety or humanity, chasing, cameras at the fore, people seeking asylum into crop fields in Europe. Even worse, one so-called journalist, writing for a small, online, right-wing publication, deliberately making a man trip, as he ran from police, her action leading to his arrest, an arrest, which you can clearly see in the footage of the event, the police officer would happily not have made (Johnson 2015). That particular action may have been an outlier, carried out by one person, strongly influenced by her individual politics. To the credit of the publication she was writing for, she was fired. Nevertheless the original thematized chasing into the fields was carried out by many—by the aptly named media pack. That moment should have been one of bearing witness. Instead a relation was set up that was not human-to-human but based on thematized "things"; the "refugee" and the "pack." It is to address such moments that Levinas's relational philosophy is needed to remind us that in each encounter we are presented with the choice of the "perpetual postponing of the hour of treason—infinitesimal difference between man and non-man" (Levinas 1979 [1969]: 35).

In contrast to the preceding, I am reminded of a surprising example described in chapter 1: the depth of understanding and humanity created by the comedians of Comic Relief. Of course there is the shared-humanity frame of laughter at work here, but I think there is something more. Those comedians needed something from their interviewees that prevented thematization; they needed their laughter and accordingly were inspired to value their alterity. Marcel Mauss argued that the unreciprocated gift wounds whoever receives it (Darnton and Kirk 2011: 60–61), but I would contend that is only half the story, for, through depriving us of the opportunity to value alterity, both parties are wounded.

To enter the borderland is to seek new knowledge, and Levinas teaches us that the most important practice on which we must focus is the building block of our profession—that face-to-face encounter of bearing witness. If we enter the relationship seeing thematized others, believing the interview is all about what *we* can offer in terms of taking up the white man's burden to report their suffering, then we cannot succeed. But if we enter seeking interlocutors, we might.

In the end, therefore, I speak only of beginnings.

Questions for the FRI Semistructured Interviews with Former DRC Refugees

Q1. How do you find out news from the Congo / Great Lakes region?

Q2. How do you find out news about other places in Africa?

Q3. How would you describe reporting of the Congo / Great Lakes region by mainstream media?

Q4. How would you describe reporting of Africa in general by the mainstream media?

Q5. I want you to think specifically about when you see refugees from sub-Saharan Africa in the media. Do you have any feelings or opinions about the way refugees are shown?

Q6. Do you feel the way refugees are shown represents you and your experiences?

Q7. Do you think the media depiction of refugee camps affects your life here in Australia?

Q8. Please watch this two-minute video from Darfur.
  - What did you think of the journalist's reporting?
  - Where there any particularly good points?
  - Where there any particularly bad points?

Q9. Please watch this two-minute video from Dadaab Refugee Camp.
  - What did you think of the journalist's reporting?
  - Where there any particularly good points?
  - Where there any particularly bad points?

Q10. Please watch this two-minute video from the DRC-Uganda border.
  - What did you think of the journalist's reporting?
  - Where there any particularly good points?
  - Where there any particularly bad points?

Q11. Thinking specifically about the style of all three videos, do you have a preference? Can you tell me why? How did that one make you feel compared to the others?

Q12. If you spent time in a refugee camp, were you ever interviewed by a journalist? Try to put yourself back into your way of thinking then—what would you have wanted to say?

Q13. What about if you were an Australian journalist in a refugee camp in Goma right now—what sort of topics would you want to write about and what sort of questions would you want to ask the refugees?

Q14. If a journalist said to you, "I want to do the best job possible when I interview people in a refugee camp," could you offer any advice?

Q15. Is there anything else you would like to say?

Questions on Sauti Piece Sent to "Co-design Actors"

My rationales are in italics.

1. Did you learn something new? (*Relates to the basic value of journalism.*)
2. Do you feel you have heard this story (or very similar) before? (*The point of this question was to try and elicit if the framing seemed familiar to the reader.*)
3. Do you think the story conveyed a sense of complexity? (*This question related directly to the research design.*)
4. Do you think the story conveyed a sense of chaos? (*This question related to the concern I had while writing that while I was trying to encode complexity it might be decoded as chaos.*)
5. If you answered yes to both complexity and chaos, was one frame stronger than the other? (*As above.*)
6. When thinking about the people in this story, what three words would you use to describe them? (*My concern here was to test whether the word "victim" would come out and whether there would be words chosen that convey agency.*)
7. Using your expertise as a journalist / NGO professional / former refugee / educator, was there something you particularly liked or disliked? (*This is to identify the different elements of particular value related to the story.*)
8. Any other comments? (*Reasoning as above, and also this question is partly the result of force of habit; I ask a variation of this question at the end of every interview.*)

Responses

Q1. Did you learn something new?

 A. *Ex-refugee from DRC and journalist*: Yes, communities are no longer waiting on international companies to run programs for them. It's good to hear locals taking initiatives to make their lives better.

 B. *Development expert, Action Aid*: The power of the collective and the vision is still alive—I started this with these women on February 14, 2008, and it is great to see and feel them still inspired.

 C. *History teacher and friend*: Yes—I had no idea that there were so many different types of grassroots organizations, and in fact I was surprised to read about how much political independence these women had—as in, they were campaigning without their husbands/families. I especially liked hearing about the women organizing street protests with placards (it reminded me about scenes from the early suffragettes movement and the civil rights movement in the United States).

 D. *Development expert*: Yes.

Q2. Do you feel that you have heard this story (or very similar) before?

    A. *Ex-refugee from DRC and journalist*: Yes, I have heard similar stories from people who are doing their best to make a better lives for their families under very difficult circumstances.

    B. *Development expert, Action Aid*: I have in the women's movement . . .

    C. *History teacher and friend*: I wasn't shocked to read that foreign workers has resulted in the expansion of prostitution—but I'm glad that you included this information. It's really important that we hear the "whole" story about foreign "assistance."

    D. *Development expert*: Yes.

Q3. Do you think the story conveyed a sense of complexity?

    A. *Ex-refugee from DRC and journalist*: Yes, it's not easy for people who have no resources to get together and run a successful and sophisticated campaign like the ladies described in the story have.

    B. *Development expert, Action Aid*: Yes it has tried.

    C. *History teacher and friend*: Yes—definitely and as a result I want to know more. . . . For example—what were the key messages on the placards? How exactly do Sauti help the women in the agricultural areas? Why did Action Aid cut their financial assistance? What long-term impact will Peace Day at Goma airport have on raising international awareness—how can these Sauti women grab onto the social hype of Peace Day and use it for their own campaign? The part about talking within the community regarding the rebels is a bit hectic. Are these women not putting themselves in danger? I am a bit confused about this.

    D. *Development Expert*: Yes.

Q4. Do you think the story conveyed a sense of chaos?

    A. *Ex-refugee from DRC and journalist*: Yes in the sense that members of the ladies group had to resort to extreme measures to get their voices heard, but like many movements, people have to resort to unorthodox measures when their welfare is at stake.

    B. *Development expert, Action Aid*: Yes it does . . .

    C. *History teacher and friend*: The story definitely alludes to the fact that these women have led difficult lives and they are trying to advocate for change in a very difficult political and social situation. However, the overarching theme of the story was that these women are determined to try; they are willing to go to great lengths such as "booking their flights" or "placard protest" based upon their own initiatives.

    D. *Development Expert*: Yes.

Q5. If you answered yes to both complexity and chaos, was one frame stronger than the other?

    A. *Ex-refugee from DRC and journalist*: No, I think both were equal.

    B. *Development expert, Action Aid*: The complexity.

    C. *History teacher and friend*: The complexities are definitely stronger than the chaos. The story focuses on the present and future actions of these women, rather than on the difficulties and injustices of their past.

    D. *Development expert*: Complexity stronger than chaos.

Q6. When thinking about the people in this story, what three words would you use to describe them?

A. *Ex-refugee from DRC and journalist*: Courageous, resilient, and inspirational.

B. *Development expert, Action Aid*: Determination, leadership, and agency!

C. *History teacher and friend*: Courageous, determined, orators.

D. *Development expert*: Resilient, determined, proud.

Q7. Using your expertise, was there something you particularly liked or disliked?

A. *Ex-refugee from DRC and journalist*: I liked the narration of the ladies' stories. It would have been nice to hear more of their experiences as well as the views of some of the men in their lives, be they husbands, partners, or clients. We often hear the ladies' side of the story, but like in most situations, there are other parties involved in the stories.

B. *Development expert, Action Aid*: I disliked the term "prostitute." As a feminist who has worked with diverse women, the terminology they prefer for their identity is sex workers and that is what should be used.

C. *History teacher and friend*: It would be better to briefly set the context at the start of the article. I don't know anything about the war in DRC . . . so a few sentences setting the scene could be useful. Dates for the war/conflicts.

D. *Development expert*: I liked that the women were not talked about in terms of victims, but it also acknowledged the hardship which they had experienced. It talked about their individual experiences, but also talked about the broader social, political, and economic contexts in which the women were situated. Loved the focus on women and women's education. Liked that they were advocacy-orientated and [had a] governance perspective, i.e., peak body and connected politically internally and also region. Like the mention of Action Aid support and also that funding had stopped but would have liked further explanation of why funding stopped.

Q8. Any other comments?

A. *Ex-refugee from DRC and journalist*: Great work, hope to read more of your reports from the field. I do appreciate it may be difficult, but try and include both sides of the story.

B. *Development expert, Action Aid*: You have made tears roll down my face. I am touched! I can connect. I know them, I feel them, and I hear them.

C. *History teacher and friend*: After reading this story I felt like I had been educated and consequently now felt "informed" about the power not just of women, but of grassroots organizations. The story encouraged me to research and find out more!

D. *Development expert*: Loved it!

I sent the story to the Action Aid development expert after sending the original tranche of questionnaires out, and so I took the opportunity to add a further question:

Q9. Do you feel that this type of story is often told by journalists reporting from the DRC?

No, they only talk about power struggle. This is a story of women's agency to transform conflict. It is a different story, a different voice.

Notes

Chapter 1

1. James is not his real name.
2. The year 2008 was chosen as the starting point of analysis because the global financial crisis shifted the news agenda around the world. Economic relations are supposed to be a key determinant of whether a country receives coverage by western media (Besova and Cooley 2009; Wu cited in Franks 2010), and Rothmyer has suggested "sustained economic progress" for sub-Saharan African countries should lead to a different type of coverage (2011: 20). My analysis was far too small to draw any inferences, with only six programs on Africa in 2008 and three programs (one a two-part special) in 2013. The end date of 2013 was chosen simply because my analysis needed to finish before my own reporting trip to the DRC in 2014.
3. For a fuller exposition see Giotis (2017).
4. This shameful history is shared with Australia's colonization, where the land was legally declared terra nullius meaning that it was deemed to contain no human inhabitants of consequence for land ownership, despite the clear presence of Aboriginal communities.
5. In my frame analysis of *Foreign Correspondent*, of the two stories based in the DRC in the period from the beginning of 2008 to the end of 2013, one was in 2009 and one was in 2010. *Both* were framed around the plight of great apes because of the ongoing conflict in the east of the country. Indeed, the reporter of these DRC stories, Eric Campbell, admits to camera: "It says much about our western sensibilities that the plight of gorillas has done more to raise awareness of the war than the plight of people" ("The Congo Connection" 2009: @24m55s). However, this knowledge doesn't stop him from framing the story as being about gorillas. In the opening Campbell says: "They [rebel groups] have slaughtered the wildlife and terrorised the local people" ("The Congo Connection" 2009: @03m30s)—not only putting people after wildlife but *ignoring the fatalities on the human side.*
6. The exposure of Stanley is recounted in Lindqvist (1998: 40–43).
7. Powers (2018: 168) argues that the self-reflection inspired by the acknowledgment of the transactional frame has not resulted in change for INGOs because the institutional conditions in which they operate make dramatic change difficult.

Chapter 2

1. I am grateful to my colleague Cale Bain for our work together on a theory and tool called the Frame Feedback Loop (FFL) (Bain and Giotis 2014). The theory helps journalists understand how an individual story frame relates to mediatized discourses, and how these mediatized discourses relate to powerful discourses in society. Finally it helps journalists consider the role of their own primary framework in choosing any particular frame.
2. All names are pseudonyms.
3. This is a technical college.
4. The traditional breaking of the fast dinner that occurs during Ramadan.
5. The rewritten article is available at https://chrisanthigiotis.wordpress.com/2014/09/23/women-of-congo-unite/.
6. The word *metaphora* is still used in modern Greek and means "to transfer."
7. *Crikey* is particularly read by journalists thanks to its dedicated reporter focusing on the media and because of its quality political reporting.
8. https://www.crikey.com.au/2015/04/21/what-it-takes-to-survive-as-an-orphan-in-a-congolese-refugee-camp/.
9. Original hyperlink in article: https://www.mercycorps.org/sites/default/files/2020-02/MercyCorps_DRC_AssessingHumanitarianResponseNorthKivu_2014.pdf.
10. Original hyperlink in the article: http://www.unhcr.org/pages/49e45c366.html.
11. Original hyperlink in the article: https://www.mercycorps.org/sites/default/files/2020-02/MercyCorps_DRC_AssessingHumanitarianResponseNorthKivu_2014.pdf.
12. Original hyperlink in the article: https://www.thestar.com.my/lifestyle/features/2014/07/07/dr-congos-tshukudu-the-allpurpose-transport-scooter/.
13. Original hyperlink in the article: https://www.youtube.com/watch?v=xdceUg4_rds.
14. Original hyperlink in the article: http://www.crisisgroup.org/en/regions/op-eds/2015/guehenno-10-wars-to-watch-in-2015.aspx.

Chapter 3

1. I refer to these humanitarian territories as Aid/Peacelands. The term "Aidland" comes from Mosse (2011a) and "Peaceland" Autesserre (2014). The paradoxical term "Peaceland" may be technically more appropriate for Goma, and indeed Autesserre's work is based on the international community working to create "peace" in the DRC. However, with such a large contingent of INGOs in Goma, the literature on Aidlands is also highly relevant, which is why I use the hybrid term.
2. I refer to international aid organizations as INGOs and local aid organizations as NGOs. Accurate figures for the number of INGOs in the world are difficult to come by, but a common estimate is 40,000. The number of local NGOs is in the millions.

It has been noted that INGOs have grown exponentially since the 1990s, and large INGOs now have budgets bigger than some OECD countries.

3. Consider the line in the Furaha article "She like everyone else answers 'the hunger.' " Also consider how the refugee camp and its various mechanisms featured heavily in the Frame Reflection Interviews in Sydney. The everyday injustices meted out in those places still sat heavily with my interviewees.

4. A term used in much of central-east and southern Africa for white person.

5. Pseudonym.

6. Pseudonym.

7. Pseudonym for one of my interviewees in the Frame Reflection Interviews—see chapter 2.

8. There are foreign correspondents who have set themselves the personal challenge of this task, such as Linda Polman, author of *Crisis Caravan* (2010), and Ian Birrell, http://www.ianbirrell.com/category/aid-development/.

9. See chapter 2 for an introduction to the concept of habitus.

Chapter 4

1. See introduction.

2. This concern is paralleled in literature on translation more broadly with studies finding "interpreters can become more focused on pleasing the person who hired them than on serving the people they are giving voice to" (Dawes 2013: 16).

3. Sambrook quotes the cost of maintaining an overseas bureau at $250,000 per annum "and significantly more for places like Iraq or Afghanistan" (2010: 12), whereas AUD$70,000 was quoted for a parachuting trip by an Australian television news crew that resulted in four stories (Willis 2016). Meanwhile stringers reporting from Syria are paid a mere $70 (Borri 2013).

Chapter 5

1. These dynamics of simplification and sedimentation are not new. In her excellent analysis of the factors influencing the reporting of the Ethiopian crisis of the mid-1980s, Suzanne Franks details how the politics of war underlying the famine was featured in the original Michael Buerk BBC documentary that turned the world's attention to Ethiopia. However, subsequent reporting of the famine by journalists did not pick up on this important element, and "as the story reverberated around the media it was picked up in an increasingly simplistic way" (Franks 2013: 101).

2. All names of frame reflection interviewees are pseudonyms.

Chapter 6

1. Stanford's itself acknowledges the importance of colonialism to its development and successful business model (Stanford's, n.d.).
2. "Shoe-leather reporting" is a term derived from the idea that journalists literally wear out their shoes hunting down unique sources of information. It is contrasted to desk journalism, where journalists rely on information coming to them (usually from powerful sources) or rely on the contacts already in their directory. Although it is not done as often as it was, it is still considered a basic virtue in the Anglo-American journalism tradition. See, for example, http://pressthink.org/2015/04/good-old-fashioned-shoe-leather-reporting/.
3. The Dutch version of *Congo: The Epic History of a People* was published in 2010. However, the English version, published in 2014, is the one I have been referencing; hence the discrepancy in dates.
4. In November 2013 the *Jakarta Post* revealed that four of six attempts by Australian authorities to turn boats around to Indonesia had been refused. Opposition leader Bill Shorten quipped at the time, "We are finding out more about what the Australian government's doing in the *Jakarta Post* than we're finding out from the government ministers" (AAP 2013: online).
5. Discussed in chapter 4.
6. This is not a critique of campaigning journalism per se. Such journalism, based on careful investigation, plays an important role in efforts to right injustices. What I am critiquing is campaigning used as a default position or style of reportage in foreign correspondence.
7. Edward Behr's well-known foreign correspondent autobiography was titled *Anyone Here Been Raped and Speaks English?* It recounts his witnessing a BBC reporter, striding among Belgian refugees waiting to be airlifted out of the Congo, who shouted, "Anyone here been raped and speaks English?" (Behr 1978: 134). The saying has become a shortcut in the profession for describing the type of reporting recounted by Eriksson Baaz and Stern (2013).

Conclusion

1. Bourdieu refers to "lifestyle" as "the best example of the unity of human behaviour of a person, but also of a group" (2005a: 44) and lifestyles are an obvious contribution to the third column of Harvey's matrix; "lived space."

Bibliography

AAP. 2013. "Boat Backdown Leaves Coalition Foreign Policy in Disarray, Says Labor." *Guardian Australia*, November 10, 2013. https://www.theguardian.com/world/2013/nov/10/boat-backdown-coalition-policy-disarray-indonesia-labor.

Aarseth, Helene, Lynne Layton, and Harriet Bjerrum Nielsen. 2016. "Conflicts in the Habitus: The Emotional Work of Becoming Modern." *Sociological Review* 64: 148–65.

ABC. 2016. "State of Fear." *Four Corners*, March 29, 2016.

ABC. 2021. "ABC Achieves Record News Audience." Accessed September 23, 2021. https://about.abc.net.au/press-releases/abc-achieves-record-news-audiences/

Achebe, Chinua. 1977. "An Image of Africa." *Massachusetts Review* 18: 782–94.

Adiche, Chimamanda. 2009. "The Danger of a Single Story." TED Talks. Accessed February 14, 2013. http://www.ted.com/talks/chimamanda_adichie_the_danger_of_a_single_story.html.

Agier, Michel. 2016. *Borderlands: Towards an Anthropology of the Cosmopolitan Condition*. London: Polity.

Agier, Michel. 2019. "Camps, Encampments, and Occupations: From the Heterotopia to the Urban Subject." *Ethnos* 84: 14–26.

Al Jazeera. 2021. "DR Congo: Dozens Detained in Beni during Anti-UN Protests." *Al Jazeera*, April 8, 2021.

Allan, Stuart, and Barbie Zelizer. 2004. "Rules of Engagement: Journalism and War." In Stuart Allan and Barbie Zelizer (eds.), *Reporting War: Journalism in Wartime*, 3–21. London: Routledge.

Allen, Tim. 1999. "Perceiving Contemporary Wars." In Tim Allen and Jean Seaton (eds.), *The Media of Conflict: War Reporting and Representations of Ethnic Violence*, 11–42. London: Zed.

Allen, Tim, and Koen Vlassenroot. 2010. "Introduction." In Tim Allen and Koen Vlassenroot (eds.), *The Lord's Resistance Army: Myth and Reality*, 1–21. London: Zed.

Alphamin Resources. 2016. "Adding Value." Accessed August 7, 2016. http://alphaminresources.com/adding-value/.

Anderson, Ben. 2012. "Affect and Biopower: Towards a Politics of Life." *Transactions of the Institute of British Geographers* 37: 28–43.

Anderson, Benedict. 1983. *Imagined Communities: Reflections on the Origin and Spread of Nationalism*. London: Verso.

Andersson, Ruben. 2018. "Profits and Predation in the Human Bioeconomy." *Public Culture* 30: 413–39.

Andersson, Ruben. 2020. "Bamako, Mali: Danger and the Divided Geography of International Intervention." In Mary Kaldor and Saskia Sassen (eds.), *Cities at War: Global Insecurity and Urban Resistance*, 25–52. New York: Columbia University Press.

Appadurai, Arjun. 1990. "Disjuncture and Difference in the Global Cultural Economy." *Theory, Culture & Society* 7: 295–310.

Appadurai, Arjun. 2006. *Fear of Small Numbers: An Essay on the Geography of Anger*. Durham, NC: Duke University Press.

Apthorpe, Raymond. 2011. "Coda. With Alice in Aidland: A Seriously Satirical Allegory." In David Mosse (ed.), *Adventures in Aidland: The Anthropology of Professionals in International Development*, 199–219. New York: Berghahn Books.

Aradau, Claudia, and Martina Tazzioli. 2019. "Biopolitics Multiple: Migration, Extraction, Subtraction." *Millennium* 48: 198–220.

Aristotle, and George A. Kennedy. 2007. *Aristotle: On Rhetoric, a Theory of Civic Discourse*, translated and edited by George A. Kennedy. Oxford: Oxford University Press.

Armoudian, Maria. 2016. *Reporting from the Danger Zone: Frontline Journalists, Their Jobs, and an Increasingly Perilous Future*. London: Routledge.

Arnsperger, Christian, and Yanis Varoufakis. 2003. "Toward a Theory of Solidarity." *Erkenntnis* 59: 157–88.

Atkinson, Philippa. 1999. "Deconstructing Media Mythologies of Ethnic War in Liberia." In Tim Allen and Jean Seaton (eds.), *The Media of Conflict: War Reporting and Representations of Ethnic Violence*, 192–218. London: Zed.

Autesserre, Séverine. 2012. "Dangerous Tales: Dominant Narratives on the Congo and Their Unintended Consequences." *African Affairs* 111: 202–22.

Autesserre, Séverine. 2014. *Peaceland: Conflict Resolution and the Everyday Politics of International Intervention*. New York: Cambridge University Press.

Ayres, Chris. 2006. *War Reporting for Cowards*. London: John Murray.

Bafilemba, F., T. Mueller, and S. Lezhnev. 2014. "The Impact of Dodd-Frank and Conflict Minerals Reforms on Eastern Congo's Conflict." The Enough Project. http://www.enoughproject.org/files/Enough%20Project%20-%20The%20Impact%20of%20Dodd-Frank%20and%20Conflict%20Minerals%20Reforms%20on%20Eastern%20Congo's%20Conflict%2010June2014.pdf.

Bain, Cale, and Chrisanthi Giotis. 2014. "Framing, Frame Analysis and Discourse in Journalism and Society." Presentation to *Toward 2020: New Directions in Journalism Education*. Toronto: Ryerson University.

Baker, Ian. 1980. "The Gatekeeper Chain: A Two-Step Analysis of How Journalists Acquire and Apply Organizational News Priorities." In Patricia Edgar (ed.), *The News in Focus: The Journalism of Exception*, 136–58. South Melbourne, Victoria: Macmillan.

Bakumanya, Bienvenu-Marie, and Samir Tounsi. 2020. "DRC Illegal Mining: The Dangerous First Link in Gold Supply Chain." *Yahoo! Finance*, September 15, 2020. https://au.finance.yahoo.com/news/drc-illegal-mining-dangerous-first-192041471.html.

Baloch, Kiyya, and Kenneth Andresen. 2020. "Reporting in Conflict Zones in Pakistan: Risks and Challenges for Fixers." *Media and Communication* 8: 37–46.

Barry-Shaw, Nikolas, Yves Engler, and Dru Oja Jay. 2012. *Paved with Good Intentions: Canada's Development NGOs from Idealism to Imperialism*. Winnipeg: Fernwood.

Beck, Ulrich. 2009. *World at Risk*. Cambridge: Polity.

Behr, Edward. 1978. *Anyone Here Been Raped and Speaks English? A Foreign Correspondent's Life behind the Lines*. London: H. Hamilton.

Belloni, Milena. 2014. "It's April, Which Means Eritrea's Refugees Are Headed North." *Global Post*. April 5, 2014. http://www.globalpost.com/dispatch/news/regions/africa/140403/its-april-which-means-eritreas-refugees-are-headed-north.

Benjamin, Walter. 1968. *Illuminations*. Ed. Hannah Arendt, trans. Harry Zohn. New York: Harcourt, Brace & World.

Benthall, Jonathan. 2010. *Disasters, Relief and the Media*. Wantage, UK: Sean Kingston Publishing.

Berglez, Peter. 2013. *Global Journalism: Theory and Practice*. New York: Peter Lang.

Bergo, Bettina. 2015. "Emmanuel Levinas." In Edward N. Zalta (ed.), *The Stanford Encyclopedia of Philosophy*. http://plato.stanford.edu/archives/sum2015/entries/levinas/.

Bernasconi, Robert. 2012. "Levinas's Ethical Critique of Levinasian Ethics." In Scott Davidson and Diane Perpich (eds.), *Totality and Infinity at 50*, 253–69. Pittsburgh, PA: Duquesne University Press.

Besova, Asya A., and Skye Chance Cooley. 2009. "Foreign News and Public Opinion: Attribute Agenda-Setting Theory Revisited." *Ecquid Novi* 30: 219–42.

Bhabha, Homi. 2004. "Foreword: Framing Fanon." In *The Wretched of the Earth*, trans. Richard Philcox, vii–xli. New York: Grove Press.

Blöbaum, Bernd. 2014. "Trust and Journalism in a Digital Environment." Reuters Institute for the Study of Journalism, University of Oxford. https://reutersinstitute.politics.ox.ac.uk/sites/default/files/2017-11/Trust%20and%20Journalism%20in%20a%20Digital%20Environment.pdf.

Borri, Francesca. 2013. "Woman's Work: The Twisted Reality of an Italian Freelancer in Syria." *Columbia Journalism Review* 52: 16–18.

Bourdieu, Pierre. 2001. "Uniting to Better Dominate." *Items and Issues* 2: 1–8.

Bourdieu, Pierre. 2005a. "Habitus." In Jean Hillier and Emma Rooksby (eds.), *Habitus: A Sense of Place*, 2nd edition, 43–49. London: Routledge.

Bourdieu, Pierre. 2005b. "The Political Field, the Social Science Field, and the Journalistic Field." In Erik Neveu and Rodney Benson (eds.), *Bourdieu and the Journalistic Field*, 29–47. Oxford: Polity.

Bourdieu, Pierre, and Terry Eagleton. 1992. "Doxa and Common Life." *New Left Review* 1: 111–21.

Boyd-Barrett, Oliver. 2004. "Understanding: The Second Casualty." In Stuart Allan and Barbie Zelizer (eds.), *Reporting War: Journalism in Wartime*, 25–42. New York: Routledge.

Brantlinger, Patrick. 1988. *Rule of Darkness: British Literature and Imperialism, 1830–1914*. Ithaca, NY: Cornell University Press.

Braudel, Fernand. 1981. *The Structures of Everyday Life: The Limits of the Possible*. Trans. Miriam Kochan, revised by Siân Reynolds. London: Collins.

Buerk, Michael. 2004. *The Road Taken*. London: Hutchinson.

Bunce, Mel. 2015. "International News and the Image of Africa: New Storytellers, New Narratives." In Julia Gallagher (ed.), *Images of Africa: Creation, Negotiation and Subversion*, 42–62. Manchester: Manchester University Press.

Büscher, Karen. 2020. "Violent Conflict and Urbanization in Eastern Democratic Republic of the Congo: The City as Safe Haven." In Mary Kaldor and Saskia Sassen (eds.), *Cities at War: Global Insecurity and Urban Resistance*, 160–83. New York: Columbia University Press.

Butler, Judith. 2016. *Frames of War: When Is Life Grievable?* New York: Verso Books.

Cain, Kenneth, Heidi Postlewait, and Andrew Thomson. 2006. *Emergency Sex (and Other Desperate Measures)*. Reading: Ebury Press.

Carey, James W. 1987. "The Press and the Public Discourse." *Center Magazine* 20: 4–16.

Carruthers, Susan L. 2004. "Tribalism and Tribulation." In Stuart Allan and Barbie Zelizer (eds.), *Reporting War: Journalism in Wartime*, 155–73. London: Routledge.

Chandler, Jo. 2013. Panel presentation, "Futures Ideas for Fast-Paced Media in a Complex World." Development Futures Conference, University of Technology Sydney.

Chouliaraki, Lilie. 2006. *The Spectatorship of Suffering*. London: Sage.

Chouliaraki, Lilie. 2013. "Re-mediation, Inter-mediation, Trans-mediation." *Journalism Studies* 14: 267–83.

Coetzee, C. 2013. "Sihle Khumalo, Cape to Cairo, and Questions of Intertextuality: How to Write about Africa, How to Read about Africa." *Research in African Literatures* 44: 62–75.

"Conflicted: The Fight over Congo's Minerals." 2015. *Fault Lines*. Television program, *Al Jazeera*. November 19, 2015.

"The Congo Connection." 2009. *Foreign Correspondent*. Television program, ABC-TV, Sydney. September 8, 2009.

Conley, David, and Stephen Lamble. 2006. *The Daily Miracle: An Introduction to Journalism*. New York: Oxford University Press.

Cooper, Hannah. 2014. "More Harm Than Good? UN's Islands of Stability in DRC." In *Oxfam Policy & Practice Blog*. May 8, 2014. Accessed March 18, 2021. https://views-voi ces.oxfam.org.uk/2014/05/islands-of-stability-in-drc/

Cottle, Simon. 2008. "Journalism and Globalization." In Karin Wahl-Jorgensen and Thomas Hanitzsch (eds.), *The Handbook of Journalism Studies*, 341–56. New York: Routledge.

Cottle, Simon. 2009a. "Global Crises in the News: Staging New Wars, Disasters and Climate Change." *International Journal of Communication* 3: 494–516.

Cottle, Simon. 2009b. *Global Crisis Reporting*. Maidenhead: Open University Press.

Cottle, Simon. 2010. "Global Crises and World News Ecology." In Stuart Allan (ed.), *The Routledge Companion to News and Journalism*, 473–84. New York: Routledge.

Cottle, Simon. 2013. "Journalists Witnessing Disaster: From the Calculus of Death to the Injunction to Care." *Journalism Studies* 14: 232–48.

Cottle, Simon. 2016. "Reporting from Unruly, Uncivil Places: Journalist Voices from the Front Line." In Simon Cottle, Richard Sambrook, and Nick Mosdell (eds.), *Reporting Dangerously*, 111–44. New York: Springer.

Cottle, Simon, and Mugdha Rai. 2006. "Between Display and Deliberation: Analyzing TV News as Communicative Architecture." *Media, Culture & Society* 28: 163–89.

Cottle, Simon, Richard Sambrook, and Nick Mosdell. 2016. *Reporting Dangerously: Journalist Killings, Intimidation and Security*. New York: Springer.

Coulter, Paddy. 1989. "Pretty as a Picture." *New Internationalist* April 1989: 10–12.

Critchley, Simon. 2002. "Introduction." In Simon Critchley and Robert Bernasconi (eds.), *The Cambridge Companion to Levinas*, 1–32. New York: Cambridge University Press.

Darnton, Andrew, and Martin Kirk. 2011. *"Finding Frames: New Ways to Engage the UK Public in Global Poverty."* London: Oxfam & Department for International Development.

Davidson, B. 1992. *The Black Man's Burden*. New York: Random House.

Davidson, Scott, and Diane Perpich. 2012. "On a Book in Midlife Crisis." In Scott Davidson and Diane Perpich (eds.), *Totality and Infinity at 50*, 1–20. Pittsburgh, PA: Duquesne University Press.

Davies, Kayt. 2012. "Safety vs Credibility: West Papua Media and the Challenge of Protecting Sources in Dangerous Places." *Pacific Journalism Review* 18: 69–82.

Davis, Dennis K., and Kurt Kent. 2013. "Journalism Ethics in a Global Communication Era: The Framing Journalism Perspective." *China Media Research* 9: 71–82.

Davis, Rebecca. 2015. "It's Official: #RhodesWillFall." *Daily Maverick*, April 9, 2015. https://www.dailymaverick.co.za/article/2015-04-09-its-official-rhodeswillfall/#.Vh7l nRYVR4M.

Dawes, James. 2013. *Evil Men*. Cambridge, MA: Harvard University Press.

de Botton, Alain. 2014. *The News: A User's Manual*. London: Penguin Books.

de Waal, Alex. 1990. *Starving in Silence: A Report on Famine and Censorship*. London: Article 19.

Dei, George J. Sefa. 2006. "Language, Race and Anti-racism: Making Important Connections." In Nuzhat Amin, George J. Sefa Dei, and Meredith Lordan (eds.), *The Poetics of Anti-racism*, 24–30. Halifax, NS: Fernwood.

Derrida, Jacques. 1999. *Adieu to Emmanuel Levinas*. Stanford, CA: Stanford University Press.

Dewey, Caitlin. 2013. "Kenyans Mock Foreign Media Coverage on Twitter." *Washington Post*, March 4, 2013. https://www.washingtonpost.com/news/worldviews/wp/2013/03/04/kenyans-mock-foreign-media-coverage-on-twitter/.

Doom, Ruddy, and Koen Vlassenroot. 1999. "Kony's Message: A New Koine? The Lord's Resistance Army in Northern Uganda." *African Affairs* 98: 5–36.

Dowden, R. 2009. *Africa: Altered States, Ordinary Miracles*. London: Portobello Books.

Dreher, Tanja. 2009. "Eavesdropping with Permission." *Borderlands* 8: 1–21.

Duffield, Mark R. 2007. *Development, Security and Unending War: Governing the World of Peoples*. Cambridge: Polity.

Duffield, Mark R. 2012. "Challenging Environments: Danger, Resilience and the Aid Industry." *Security Dialogue* 43: 475–92.

Duffield, Mark R. 2013. "How Did We Become Unprepared? Emergency and Resilience in an Uncertain World." *British Academy Review* 21: 55–58.

Duffield, Mark R. 2014. "From Immersion to Simulation: Remote Methodologies and the Decline of Area Studies." *Review of African Political Economy* 41: S75–S94.

Dunn, Kevin C. 2003. *Imagining the Congo*. New York: Palgrave Macmillan.

Easterly, William Russell. 2006. *The White Man's Burden: Why the West's Efforts to Aid the Rest Have Done So Much Ill and So Little Good*. New York: Penguin Press.

"Edelman Trust Barometer." 2020. Accessed July 21, 2020. https://www.edelman.com.au/research/edelman-trust-barometer-2020.

Egeland, Jan, Adele Harmer, and Abby Stoddard. 2011. "To Stay and Deliver: Good Practice for Humanitarians in Complex Security Environments." United Nations: Office for the Coordination of Humanitarian Affairs.

Ekström, Mats. 2002. "Epistemologies of TV Journalism: A Theoretical Framework." *Journalism* 3: 259–82.

Entman, Robert M. 1991. "Framing US Coverage of International News: Contrasts in Narratives of the KAL and Iran Air Incidents." *Journal of Communication* 41: 6–27.

Entman, Robert M. 1993. "Framing: Toward Clarification of a Fractured Paradigm." *Journal of Communication* 43: 51–58.

Erickson, Emily, and John Maxwell Hamilton. 2006. "Foreign Reporting Enhanced by Parachute Journalism." *Newspaper Research Journal* 27: 33–47.

Ericson, Richard, Patricia Baranek, and Janet Chan. 1989. "Negotiating Control." In Richard Ericson, Patricia Baranek, and Janet Chan (eds.), *Negotiating Control: A Study of News Sources*, 377–98. Toronto: University of Toronto Press.

Eriksson Baaz, Maria, and Maria Stern. 2013. *Sexual Violence as a Weapon of War? Perceptions, Prescriptions, Problems in the Congo and Beyond*. New York: Zed.

Eyben, Rosalind. 2011. "The Sociality of International Aid and Policy Convergence." In David Mosse (ed.), *Adventures in Aidland: The Anthropology of Professionals in International Development*, 151–60. New York: Berghahn Books.

Ezeru, Chikaire Wilfred Williams. 2021. "The Continued Domination of Western Journalists in Global African News Telling: The Imperatives and Implications." *African Journalism Studies* 42: 36–55.

Fahey, Daniel. 2009. "'Congo Gold': Three Problems with the *60 Minutes* Story." Accessed September 4, 2014. http://africanarguments.org/2009/12/11/three-problems-with-the-60-minutes/.

Fahey, Daniel, and Bally Mutumayi. 2019. "The Transition from Artisanal to Industrial Mining at Bisie, Democratic Republic of the Congo." Accessed April 23, 2022. https://www.researchgate.net/publication/332877943_The_Transition_from_Artisanal_to_Industrial_Mining_at_Bisie_Democratic_Republic_of_the_Congo.

Fanon, Frantz. 1986 [1952]. *Black Skin, White Masks*. Trans. Richard Philcox. London: Pluto Press.

Fanon, Frantz. 2004 [1963]. *The Wretched of the Earth*. Trans. Richard Philcox. New York: Grove Press.

Fisk, Robert. 2005. "Hotel Journalism Gives American Troops a Free Hand as the Press Shelters Indoors." *The Independent*. January 17, 2005.

Fiske, Lucy. 2016. *Human Rights, Refugee Protest and Immigration Detention*. London: Palgrave Macmillan.

Franks, Suzanne. 2010. "The Neglect of Africa and the Power of Aid." *International Communication Gazette* 72: 71–84.

Franks, Suzanne. 2013. *Reporting Disasters*. London: C. Hurst.

French, Howard. 2005. *A Continent for the Taking*. New York: Vintage Books.

French, Howard. 2011. "Congo: Rape, Savagery, and Stereotypes, the Heart of Darkness." Crisis in the Congo. Accessed February 25, 2015. https://www.youtube.com/watch?v=NXJEVoaHoHU.

Friedman, Sam. 2015. "Habitus Clivé and the Emotional Imprint of Social Mobility." *Sociological Review* 64: 129–47.

Garnham, Nicholas, and Raymond Williams. 1980. "Pierre Bourdieu and the Sociology of Culture: An Introduction." *Media, Culture & Society* 2: 209–23.

"The General's Dilemma." 2008. *Foreign Correspondent*. Television program, ABC-TV, Sydney. October 28, 2008.

Georgieva, Kristalina. 2021. "The Great Divergence: A Fork in the Road for the Global Economy." *IMFBlog*, February 24, 2021. https://blogs.imf.org/2021/02/24/the-great-divergence-a-fork-in-the-road-for-the-global-economy/.

Ghani, Bilquis, and Lucy Fiske. 2020. "'Art Is My Language': Afghan Cultural Production Challenging Islamophobic Stereotypes." *Journal of Sociology* 56: 115–29.

Giotis, Chrisanthi. 2011. "South Sudan's Tomorrow People Think beyond Their Squalor." It Began in Africa, July 29, 2011. http://www.itbeganinafrica.com/inspiring-people/south-sudan%E2%80%99s-tomorrow-people-think-beyond-their-squalor.

Giotis, Chrisanthi. 2015. "Crikey—Something Published on DRC Refugees." *A Journo—on Journos—in Africa*, April 24, 2015. https://chrisanthigiotis.wordpress.com/2015/04/24/crikey-something-published-on-drc-refugees/.

Giotis, Chrisanthi. 2017. "Not Just a Victim of War." Doctoral diss., University of Technology Sydney.

Giotis, Chrisanthi. 2019a. "More Than a Victim: Thinking through Foreign Correspondents' Representations of Women in Conflict." In Rita Shackel and Lucy Fiske (eds.), *Rethinking Transitional Gender Justice*, 97–117. New York: Palgrave Macmillan.

Giotis, Chrisanthi. 2019b. "'Stop Playing Politics': Refugees Stuck in Indonesia Rally against UNHCR for Chronic Waiting." *The Conversation*, October 8, 2019.

Giotis, Chrisanthi. 2021. "Dismantling the Deadlock: Australian Muslim Women's Fightback against the Rise of Right-Wing Media." *Social Sciences* 10: 71.

Giotis, Chrisanthi, and Christopher Hall. 2021. "Better Foreign Correspondence Starts at Home: Changing Practice through Diasporic Knowledge." *Journalism Practice*: 1–18. https://doi.org/10.1080/17512786.2021.1930105.

"Go Back to Where You Came from S3 E2." 2015. *Go Back to Where You Came From*. Television program, SBS TV, Sydney. July 29, 2015.

Go, Julian. 2013. "Fanon's Postcolonial Cosmopolitanism." *European Journal of Social Theory* 16: 208–25.

Goffman, Erving. 1975. *Frame Analysis: An Essay on the Organization of Experience*. New York: Penguin.

Golan, Guy. 2006. "Inter-media Agenda Setting and Global News Coverage: Assessing the Influence of the *New York Times* on Three Network Television Evening News Programs." *Journalism Studies* 7: 323–33.

Goldstein, Tom. 2007. *Journalism and Truth: Strange Bedfellows*. Evanston, IL: Northwestern University Press.

Gordimer, Nadine. 1985. "The Essential Gesture: Writers and Responsibility." *Granta* 15: 137–51.

Granqvist, Raoul J. 2012. "Photojournalism's White Mythologies: Eliot Elisofon and *Life* in Africa, 1959–1961." *Research in African Literatures* 43: 84–105.

Greco, John. 1999. "Introduction: What Is Epistemology?" In Ernest Sosa and John Greco (eds.), *The Blackwell Guide to Epistemology*, 1–31. Malden, MA: Blackwell.

Gregg, Melissa, and Gregory J. Seigworth. 2010. "An Inventory of Shimmers." In Melissa Gregg and Gregory J. Seigworth (eds.), *The Affect Theory Reader*, 1–25. Durham, NC: Duke University Press.

"Gruen XL." 2015. *Gruen Planet*. Series 7 Episode 8, Extended version on ABC iView. Television program, ABC-TV, Sydney. October 28, 2015.

Gruley, Joel, and Chris S. Duvall. 2012. "The Evolving Narrative of the Darfur Conflict as Represented in the *New York Times* and the *Washington Post*, 2003–2009." *GeoJournal* 77: 29–46.

Hall, Stuart. 1978. *Policing the Crisis: Mugging, the State, and Law and Order*. London: Macmillan.

Hall, Stuart. 2013a. "Introduction." In Stuart Hall, Jessica Evans, and Sean Nixon (eds.), *Representation*, xvii–xxvi. Milton Keynes: Sage; Open University.

Hall, Stuart. 2013b. "The Work of Representation." In Stuart Hall, Jessica Evans, and Sean Nixon (eds.), *Representation*, 1–47. Sage; Open University.

Hamilton, John Maxwell, and Eric Jenner. 2004. "Redefining Foreign Correspondence." *Journalism* 5: 301–21.

Hannerz, Ulf. 2004. *Foreign News: Exploring the World of Foreign Correspondents*. Chicago: University of Chicago Press.

Harden, Blaine. 1991. *Africa: Dispatches from a Fragile Continent.* Boston: Houghton Mifflin.

Harden, Blaine, Alan Cowell, Ian Fisher, Norimitsu Onishi, and Rachel L. Swarns. 2000. "Africa's Gems: Warfare's Best Friend." *New York Times,* April 6, 2000.

Hardt, Michael, and Antonio Negri. 2004. *Multitude: War and Democracy in the Age of Empire.* New York: Penguin.

Harper, Ian. 2011. "World Health and Nepal: Producing Internationals, Healthy Citizenship and the Cosmopolitan." In David Mosse (ed.), *Adventures in Aidland: The Anthropology of Professionals in International Development,* 123–38. New York: Berghahn Books.

Harris, Janet, and Kevin Williams. 2018. *Reporting War and Conflict.* New York: Routledge.

Harrison, Graham. 2010. "The Africanization of Poverty: A Retrospective on 'Make Poverty History.'" *African Affairs* 109: 391–408.

Harrison, Paul, and Robin Palmer. 1986. *News out of Africa: Biafra to Band Aid.* London: Hilary Shipman.

Harvey, David. 1996. *Justice, Nature and the Geography of Difference.* Cambridge: Blackwell.

Harvey, David. 2006. *Spaces of Global Capitalism.* New York: Verso.

Harvey, David. 2009. *Cosmopolitanism and the Geographies of Freedom.* New York: Columbia University Press.

Hawk, Beverly G. 1992. *Africa's Media Image.* New York: Praeger.

Heaton, Laura. 2013. "What Happened in Luvungi? On Rape and Truth in Congo." *Foreign Policy,* March–April 2013. http://www.foreignpolicy.com/articles/2013/03/04/ what_happened_in_luvungi?page=full.

Heng, Yee-Kuang. 2006. "The 'Transformation of War' Debate: Through the Looking Glass of Ulrich Beck's World Risk Society." *International Relations* 20: 69–91.

Hintzen, Percy C. 2008. "Desire and the Enrapture of Capitalist Consumption: Product Red, Africa, and the Crisis of Sustainability." *Journal of Pan African Studies* 2: 77–91.

Holmwood, John. 2013. "What Does Sociology Offer Our Society Today?" Australian Sociological Association. Accessed April 21, 2021. https://www.youtube.com/ watch?v=e4zh5dItns4&t=1290s.

Hunter-Gault, Charlayne. 2006. *New News out of Africa: Uncovering Africa's Renaissance.* New York: Oxford University Press.

"Interview C." 2014. Fixer and stringer, September 5, 2014, Goma, DRC.

"Interview D." 2014. Journalist, stringer, and fixer, August 24, 2014, Goma, DRC.

"Interview E." 2014. NGO worker, August 26, 2014, Goma, DRC.

"Interview F." 2014. International worker, August 30, 2014, Goma, DRC.

"Interview H." 2014. Fixer and stringer, September 6, 2014, Goma, DRC.

"Interview I." 2014. Fixer and stringer, September 7, 2014, Goma, DRC.

"Interview J." 2014. Fixer, September 7, 2014, Goma, DRC.

"Interview L." 2014. Journalist, stringer, and fixer, September 12, 2014, Goma, DRC.

IRIN. 2011. "Aid Policy: Staff Security—'Bunkerization' versus Acceptance." *IRIN Humanitarian News and Analysis.* April 13, 2011.

Iyengar, Shanto. 1994. *Is Anyone Responsible? How Television Frames Political Issues.* Chicago: University of Chicago Press.

Jackson, Jeffrey T. 2005. *The Globalizers: Development Workers in Action.* Baltimore: Johns Hopkins University Press.

Johnson, Krystal. 2015. "Hungarian Camerawoman Who Tripped Refugee Plans to Sue Victim and Facebook." *Yahoo 7 News,* October 22, 2015. https://au.news.yahoo.com/a/

29875659/hungarian-camerawoman-who-tripped-refugee-plans-to-sue-victim-and-facebook/.

Johnson, Peter. 2006. "Unravelling Foucault's 'Different Spaces.'" *History of the Human Sciences* 19: 75–90.

Kaker, Sobia Ahmad. 2020. "Responding to, or Perpetuating, Urban Insecurity? Enclave-Making in Karachi." In Mary Kaldor and Saskia Sassen (eds.), *Cities at War: Global Insecurity and Urban Resistance*, 133–59. New York: Columbia University Press.

Kaldor, Mary. 2016. "How Peace Agreements Undermine the Rule of Law in New War Settings." *Global Policy* 7: 146–55.

Kaldor, Mary, and Saskia Sassen, eds. 2020. *Cities at War: Global Insecurity and Urban Resistance*. New York: Columbia University Press.

Keane, Fergal. 2004. "Trapped in a Time-Warped Narrative." *Nieman Reports* 58: 8–10.

Keeble, Richard. 2004. "Information Warfare in an Age of Hyper-militarism." In Stuart Allan and Barbie Zelizer (eds.), *Reporting War: Journalism in Wartime*, 43–58. New York: Routledge.

Kim, Hosu, and Jamie Bianco. 2007. *The Affective Turn: Theorizing the Social*. Durham, NC: Duke University Press.

Koddenbrock, Kai. 2012. "Recipes for Intervention: Western Policy Papers Imagine the Congo." *International Peacekeeping* 19: 549–64.

Kothari, Uma. 2006. "An Agenda for Thinking about 'Race' in Development." *Progress in Development Studies* 6: 9–23.

Lamar, Kendrick. 2015. "Alright." In *To Pimp a Butterfly*. Aftermath/Interscope (Top Dawg Entertainment).

Lancione, Michele. 2014. "The Spectacle of the Poor. Or: 'Wow!! Awesome. Nice to Know That People Care!'" *Social & Cultural Geography* 15: 693–713.

Laudati, Ann, and Charlotte Mertens. 2019. "Resources and Rape: Congo's (Toxic) Discursive Complex." *African Studies Review* 62: 57–82.

Lederman, Jim. 1992. *Battle Lines*. New York: Henry Holt.

Leith, Denise. 2004. *Bearing Witness*. Sydney: Random House Australia.

Levinas, Emmanuel. 1979 [1969]. *Totality and Infinity*. The Hague: Martinus Nijhoff.

Liebes, Tamar, and Zohar Kampf. 2009. "Performance Journalism: The Case of Media's Coverage of War and Terror." *Communication Review* 12: 239–49.

Lindqvist, Sven. 1998. *Exterminate All the Brutes*. London: Granta Publications.

Loewenstein, Antony. 2013. *Profits of Doom: How Vulture Capitalism Is Swallowing the World*. Carlton, Victoria: Melbourne University Publishing.

Logan, Mawuena. 2001. "The Myth of Postcolonial Africa and Juvenile Literature." *Obsidian III* 3: 126+.

Lundstrom, Marjie. 2002. "Parachute Journalism." *Poynter.org*. August 15, 2002. https://www.poynter.org/archive/2002/parachute-journalism/

Luyendijk, Joris. 2009. *Fit to Print: Misrepresenting the Middle East*. Carlton North, Victoria: Scribe Publications.

Madgwick, Steve. 2011. "The Trouble with Being Arrested in Sudan." ItbeganinAfrica.com. July 30, 2011. http://itbeganinafrica.com/inspiring-blog/trouble-being-arrested-sudan.

Madondo, Bongani. 2008. "Vanity Farce: The Africa Issue." *Transition* 98: 170–79.

Mafe, Diana Adesola. 2011. "(Mis) Imagining Africa in the New Millennium: The Constant Gardener and Blood Diamond." *Camera Obscura* 25: 69–99.

Mahon, Michael. 1992. *Foucault's Nietzschean Genealogy: Truth, Power, and the Subject.* Albany: State University of New York Press.

Mamdani, Mahmood. 1996. *Citizen and Subject: Contemporary Africa and the Legacy of Late Colonialism.* Princeton, NJ: Princeton University Press.

Mamdani, Mahmood. 2007. "The Politics of Naming: Genocide, Civil War, Insurgency." *London Review of Books* 29: 5–8. http://www.lrb.co.uk.ezproxy.lib.uts.edu.au/v29/n05/mahmood-mamdani/the-politics-of-naming-genocide-civil-war-insurgency.

Mamdani, Mahmood. 2009. "There May Have Been No Water but the Province Was Awash with Guns." *New Statesman*, 34–37. June 8, 2009.

Manderson, Desmond. 2012. "Law, Ethics, and the Unbounded Duty of Care." In Scott Davidson and Diane Perpich (eds.), *Totality and Infinity at 50*, 153–70. Pittsburgh, PA: Duquesne University Press.

Martin, Mary. 2020. "A Tale of Two Cities: Ciudad Juárez, El Paso, and Insecurity at the U.S.-Mexico Border." In Mary Kaldor and Saskia Sassen (eds.), *Cities at War: Global Insecurity and Urban Resistance*, 103–33. New York: Columbia University Press.

Mbembe, Achille. 2001. *On the Postcolony.* Berkeley: University of California Press.

Mbembe, Achille. 2003. "Necropolitics." *Public Culture* 15: 11–40.

McLaughlin, Greg. 2002. *The War Correspondent.* London: Pluto Press.

McNulty, Mel. 1999. "Media Ethnicization and the International Response to War and Genocide in Rwanda." In Tim Allen and Jean Seaton (eds.), *The Media of Conflict: War Reporting and Representations of Ethnic Violence*, 268–86. New York: Zed.

Meek, Allen. 2016. *Biopolitical Media: Catastrophe, Immunity and Bare Life.* New York: Routledge.

"Meet the Janjaweed." 2008. *Foreign Correspondent.* Television program, ABC-TV, Sydney. August 3, 2008.

Mercer, Neil, Sean Nicholls, Ellen Connolly, and Valerie Lawson. 2019 [2003]. "From the Archives: Sydney Protests the Iraq War." *Sydney Morning Herald.* February 17, 2003. https://www.smh.com.au/world/middle-east/from-the-archives-sydney-protests-the-iraq-war-20190214-p50xtd.html.

Miller, Christopher. 1985. *Blank Darkness: Africanist Discourse in French.* Chicago: University of Chicago Press.

Min, Seong Jae. 2018. *As Democracy Goes, So Does Journalism: Evolution of Journalism in Liberal, Deliberative, and Participatory Democracy.* Lanham, MD: Rowman & Littlefield.

Moeller, Susan D. 2004. "A Moral Imagination." In Stuart Allan and Barbie Zelizer (eds.), *Reporting War: Journalism in Wartime*, 59–76. London: Routledge.

Moeller, Susan D. 2006. "Regarding the Pain of Others: Media, Bias and the Coverage of International Disasters." *Journal of International Affairs* 59: 173–96.

Mohdin, Aamna. 2018. "One Person Was Displaced Every Two Seconds in 2017." Quartz, June 20, 2018. https://qz.com/1310119/world-refugee-day-one-person-was-displaced-every-two-seconds-in-2017/.

Moore, Barrington. 1972. *Reflections on the Causes of Human Misery and upon Certain Proposals to Eliminate Them.* London: Allen Lane, Penguin Press.

Mosse, David, ed. 2011a. *Adventures in Aidland: The Anthropology of Professionals in International Development.* New York: Berghahn Books.

Mosse, David. 2011b. "Introduction: The Anthropology of Expertise and Professionals in International Development." In David Mosse (ed.), *Adventures in Aidland: The Anthropology of Professionals in International Development*, 1–31. New York: Berghahn Books.

Moyo, Dambisa. 2009. *Dead Aid: Why Aid Is Not Working and How There Is a Better Way for Africa*. London: Penguin.

Mudimbe, V. Y. 1994. *The Idea of Africa*. Bloomington: Indiana University Press.

Muehlenbeck, Philip. 2012. "Who Killed Hammarskjöld? Susan Williams' Formidable Book." Accessed March 11, 2015. http://www.dag-hammarskjold.com.

Murrell, Colleen. 2009. "Fixers and Foreign Correspondents: News Production and Autonomy." *Australian Journalism Review* 31: 5–17.

Murrell, Colleen. 2013. "International Fixers: Cultural Interpreters or 'People Like Us'?" *Ethical Space* 10: 72–79.

Murrell, Colleen. 2014. "The Vulture Club: International Newsgathering via Facebook." *Australian Journalism Review* 36: 15–27.

Murrell, Colleen. 2015. *Foreign Correspondents and International Newsgathering: The Role of Fixers*. New York: Routledge.

Muspratt, Matthew, and H. Leslie Steeves. 2012. "Rejecting Erasure Tropes of Africa: The Amazing Race Episodes in Ghana Counter Postcolonial Critiques." *Communication, Culture & Critique* 5: 533–40.

Myambo, Melissa Tandiwe, and Pier Paolo Frassinelli. 2019. "Introduction: Thirty Years of Borders since Berlin." *New Global Studies* 13: 277–300.

Nash, Chris. 2013. "Journalism as a Research Discipline." *Pacific Journalism Review* 19: 123–35.

Nash, Chris. 2016. *What Is Journalism? The Art and Politics of a Rupture*. London: Palgrave Macmillan.

Ndlela, Nkosi. 2005. "The African Paradigm: The Coverage of the Zimbabwean Crisis in the Norwegian Media." *Westminster Papers in Communication and Culture*, Special Issue (November): 71–90.

Neveu, Erik, and Rodney Benson. 2005. "Introduction." In Erik Neveu and Rodney Benson (eds.), *Bourdieu and the Journalistic Field*, 1–25. Oxford: Polity.

Newbury, David. 1998. "Understanding Genocide." *African Studies Review* 41: 73–97.

Nobelstiftelsen. 2006. *Nobel Lectures: From the Literature Laureates, 1986 to 2005*. Carlton, Victoria: Melbourne University Publishing.

Nothias, Toussaint. 2014. "'Rising.' 'Hopeful.' 'New': Visualizing Africa in the Age of Globalization." *Visual Communication* 13: 323–39.

Nothias, Toussaint. 2020. "Postcolonial Reflexivity in the News Industry: The Case of Foreign Correspondents in Kenya and South Africa." *Journal of Communication* 70: 245–73.

Nyabola, Nanjala. 2014. "Why Do Western Media Get Africa Wrong?" *Al Jazeera*, January 2, 2014. http://www.aljazeera.com/indepth/opinion/2014/01/why-do-western-media-get-africa-wrong-20141152641935954.html.

O'Neill, Brendan. 2012. "Dangers of the 'Journalism of Attachment.'" *The Drum*. ABC News, February 24, 2012.

O'Neill, Deirdre, and Tony Harcup. 2008. "News Values and Selectivity." In Karin Wahl-Jorgensen and Thomas Hanitzsch (eds.), *The Handbook of Journalism Studies*, 161–74. New York: Routledge.

Paech, Michaela. 2004. "A Photograph Is Worth More Than a Thousand Words: The Impact of Photojournalism on Charitable Giving." Doctoral diss., University of London.

Palmer, Jerry, and Victoria Fontan. 2007. "'Our Ears and Our Eyes': Journalists and Fixers in Iraq." *Journalism* 8: 5–24.

Palmer, Lindsay. 2018. *Becoming the Story: War Correspondents since 9/11.* Urbana: University of Illinois Press.

Palmer, Lindsay. 2019. *The Fixers: Local News Workers and the Underground Labor of International Reporting.* New York: Oxford University Press.

Park, S., C. Fisher, K. McGuinness, J. Y. Lee, and K. McCallum. 2021. "Digital News Report: Australia 2021." Canberra: News & Media Research Centre, University of Canberra.

Patton, Michael Quinn. 2002. *Qualitative Research and Evaluation Methods.* Thousand Oaks, CA: Sage.

Peake, Linda, and Audrey Kobayashi. 2002. "Policies and Practices for an Antiracist Geography at the Millennium." *Professional Geographer* 54: 50–61.

Pedelty, Mark. 1995. *War Stories: The Culture of Foreign Correspondents.* New York: Routledge.

Perera, Suda. 2017. "Bermuda Triangulation: Embracing the Messiness of Researching in Conflict." *Journal of Intervention and Statebuilding* 11: 42–57.

Perera, Suda. 2018. "Thinking and Working Politically in Protracted Conflict: Distrust and Resistance to Change in the DRC." Developmental Leadership Program, University of Birmingham.

Perpich, Diane. 2012. "Don't Try This at Home." In Scott Davidson and Diane Perpich (eds.), *Totality and Infinity at 50*, 127–52. Pittsburgh, PA: Duquesne University Press.

Picard, Robert G. 2014. "Deficient Tutelage: Challenges of Contemporary Journalism Education." Conference Keynote. *Toward 2020: New Directions in Journalism Education.* Toronto: Ryerson Journalism Research Centre. https://ryersonjournalism.ca/2014/11/18/deficient-tutelage-challenges-of-contemporary-journalism-education/

Picard, Robert G., and Hannah Storm. 2016. *The Kidnapping of Journalists: Reporting from High-Risk Conflict Zones.* London: I.B. Tauris.

Pilger, John. 1986. "Preface." In Ben Kiernan (ed.), *Burchett Reporting the Other Side of the World, 1939–1983.* London: Quartet Books.

Plaut, Shayna, and Peter Klein. 2019. "'Fixing' the Journalist-Fixer Relationship: A Critical Look towards Developing Best Practices in Global Reporting." *Journalism Studies* 20: 1696–713.

Pole Institute. 2014. "Suspension des activités des coopératives minières à Bisie au Nord-Kivu." Accessed August 7, 2016. http://www.pole-institute.org/news/suspension-des-activités-des-coopératives-minières-à-bisie-au-nord-kivu.

Polman, Linda. 2010. *The Crisis Caravan: What's Wrong with Humanitarian Aid?* New York: Metropolitan Books.

Power, Marcus. 2006. "Anti-racism, Deconstruction and 'Overdevelopment.'" *Progress in Development Studies* 6: 24–39.

Powers, Matthew. 2015. "NGOs as Journalistic Entities: The Possibilities, Promises and Limits of Boundary Crossing." In Matt Carlson and Seth C. Lewis (eds.), *Boundaries of Journalism: Professionalism, Practices and Participation*, 186–200. New York: Routledge, Taylor & Francis Group.

Powers, Matthew. 2016. "The New Boots on the Ground: NGOs in the Changing Landscape of International News." *Journalism* 17: 401–16.

Powers, Matthew. 2018. *NGOs as Newsmakers: The Changing Landscape of International News.* New York: Columbia University Press.

Prigg, Mark. 2017. "The World at War: Stunning Interactive Map Reveals Every Conflict Currently Active around the World." Dailymail.com, April 29, 2017. https://www.dailymail.co.uk/sciencetech/article-4453666/The-world-war-Interactive-map-reveals-conflicts.html.

"PS Events: Grassroots Globalization." 2021. Project Syndicate. Accessed April 26, 2021. https://www.project-syndicate.org/event/grassroots-globalization.

Rajak, Dinah, and Jock Stirrat. 2011. "Parochial Cosmopolitanism and the Power of Nostalgia." In David Mosse (ed.), *Adventures in Aidland: The Anthropology of Professionals in International Development*, 161–76. New York: Berghahn Books.

Ramalingam, Ben. 2013. *Aid on the Edge of Chaos: Rethinking International Cooperation in a Complex World*. Oxford: Oxford University Press.

Rawlence, Ben. 2012. *Radio Congo: Signals of Hope from Africa's Deadliest War*. Oxford: Oneworld Publications.

Reese, Stephen D. 2001. "Understanding the Global Journalist: A Hierarchy-of-Influences Approach." *Journalism Studies* 2: 173–87.

Ricchiardi, Sherry. 2006. "The Limits of the Parachute." *American Journalism Review* 28: 40–47.

Robin, Myriam. 2015. "Chris Kenny, Refugee Advocate at Odds over Nauru Rape Complaint Interview." *Crikey*. October 22, 2015. crikey.com.au.

Rothmyer, Karen. 2011. "Hiding the Real Africa." *Columbia Journalism Review* 49: 18–20.

Roy Morgan. 2019. "ABC Still Most Trusted, Facebook Improves." http://www.roymorgan.com/findings/8064-abc-remains-most-trusted-media-201907220424.

RSF. 2020a. "Round Up 2020." Accessed January 18, 2020. https://rsf.org/en/news/rsfs-2020-round-50-journalists-killed-two-thirds-countries-peace.

RSF. 2020b. "Round Up 2020: Journalists Detained, Held Hostage and Missing." https://rsf.org/sites/default/files/rsfs_2020_round-up_0.pdf.

Ryle, Gerard. 2016. "How the Panama Papers Journalists Broke the Biggest Leak in History." TED Talks. Accessed February 26, 2019. https://www.ted.com/talks/gerard_ryle_how_the_panama_papers_journalists_broke_the_biggest_leak_in_history?language=en.

Said, Edward. 1993. *Culture and Imperialism*. London: Chatto & Windus.

Said, Edward. 1995. "Afterword to the 1995 Printing." In Said, *Orientalism*, 329–52. London: Penguin.

Sambrook, Richard. 2010. "Are Foreign Correspondents Redundant? The Changing Face of International News." Oxford: Reuters Institute for the Study of Journalism.

Sara, Sally. 2007. *Gogo Mama: A Journey into the Lives of 12 African Women*. Sydney: Macmillan.

Sara, Sally. 2014. "The Return: One Correspondent's Personal Experience with Trauma." *Correspondents Report*. ABC, October 12, 2014. http://www.abc.net.au/correspondents/content/2014/s4105025.htm.

Sartre, Jean-Paul. 1963. "Preface." In *The Wretched of the Earth*, trans. Richard Philcox, 7–31. New York: Grove Press.

Sassen, Saskia. 2014. *Expulsions: Brutality and Complexity in the Global Economy*. Cambridge, MA: Harvard University Press.

Saunders, Doug. 2016. "Integration: A New Strategy." *Globe and Mail* (Toronto). January, 14, 2016. http://www.theglobeandmail.com/news/world/saunders-avert-extremism-before-it-start-by-building-betterneighbourhoods/article27403775/

Scheufele, Dietram A. 2000. "Agenda-Setting, Priming, and Framing Revisited: Another Look at Cognitive Effects of Political Communication." *Mass Communication and Society* 3: 297–316.

Schön, Donald A. 1983. *The Reflective Practitioner: How Professionals Think in Action.* New York: Basic Books.

Schön, Donald A. 2018 [1989]. "Donald Schön Lecture." Special Collections Iowa State University Library. Accessed April 21, 2021. https://www.youtube.com/watch?v=Ld9QJcMiNMo.

Schön, Donald A., and Martin Rein. 1994. *Frame Reflection: Toward the Resolution of Intractable Policy Controversies.* New York: Basic Books.

Schwarz, Jon. 2015. "A Short History of US Bombing of Civilian Facilities." *The Intercept,* October 8, 2015. https://theintercept.com/2015/10/07/a-short-history-of-u-s-bombing-of-civilian-facilities/

Scott, Martin. 2016. "How Not to Write about Writing about Africa." In Melanie Bunce, Suzanne Franks, and Chris Paterson (eds.), *Africa's Media Image in the 21st Century: From the "Heart of Darkness" to "Africa Rising"*, 40–51. New York: Routledge.

Seaton, Jean. 1999. "The New 'Ethnic' Wars and the Media." In Tim Allen and Jean Seaton (eds.), *The Media of Conflict: War Reporting and Representations of Ethnic Violence*, 43–63. New York: Zed.

Serena Hotels. 2021. "Goma Serena Hotel." Accessed March 13, 2021. https://www.serenahotels.com/gomaserena/en/default.html.

Seth, Sanjay. 2014. "Whose Emergency?" Workshop at the University of Technology, Sydney, May 12, 2014.

Shineberg, Dorothy. 1986. "Wilfred Burchett's New Caledonia." In Ben Kiernan (ed.), *Burchett Reporting the Other Side of the World, 1939–1983*, 137–47. London: Quartet Books.

Silverstone, Roger. 2002. "Regulation and the Ethics of Distance: Distance and the Ethics of Regulation." In Robin E. Mansell, Amy Mahan, and Rohan Samarajiva (eds.), *Networking Knowledge for Information Societies: Institutions and Intervention*, 279–85. Delft, the Netherlands: Delft University Press.

Silverstone, Roger. 2013. *Media and Morality: On the Rise of the Mediapolis.* New York: John Wiley & Sons.

Simons, Margaret. 1999. *Fit to Print: Inside the Canberra Press Gallery.* Sydney: UNSW Press.

Smith, Hazel, and R. T. Dean. 2009. *Practice-Led Research, Research-Led Practice in the Creative Arts.* Edinburgh: Edinburgh University Press.

Stanford's. n.d. "Our History." Accessed March 29, 2019. www.stanfords.co.uk.

Sundaram, Anjan. 2014. "We're Missing the Story: The Media's Retreat from Foreign Reporting." *New York Times,* July 25, 2014.

Sundaram, Anjan. 2016. "We're Missing the Story: The Media's Retreat from Foreign Reporting." In Melanie Bunce, Suzanne Franks, and Chris Paterson (eds.), *Africa's Media Image in the 21st Century: From the "Heart of Darkness" to "Africa Rising"*, 99–101. New York: Routledge.

"The Swingers." 2010. *t.* Television program, ABC-TV, Sydney.

Tanter, Richard. 1986. "Burchett and Hiroshima." In Ben Kiernan (ed.), *Burchett Reporting the Other Side of the World, 1939–1983*, 13–40. London: Quartet Books.

Tarrow, Sidney. 2010. "Preface." In Stefaan Walgrave, Dieter Rucht, and Sidney Tarrow (eds.), *World Says No to War: Demonstrations against the War on Iraq*, vii–xi. Minneapolis: University of Minnesota Press.

Taylor, Magnus. 2012. "Radio Congo: Taking the Slow Road through Central Africa—a Review by Magnus Taylor." June 30, 2014. http://africanarguments.org/2012/09/12/radio-congo-on-a-slow-boat-through-central-africa---a-review-by-magnus-taylor/.

Taylor, Sidney, ed. 1967. *The New Africans*. London: Hamlyn.

Tewksbury, David, and Dietram Scheufele. 2009. "News Framing Theory and Research." In Jennings Bryant and Mary Beth Oliver (eds.), *Media Effects: Advances in Theory and Research*, 17–33. New York: Routledge.

Todd, Sharon. 2003. *Learning from the Other: Levinas, Psychoanalysis, and Ethical Possibilities in Education*. Albany: State University of New York Press.

Topinka, Robert J. 2010. "Foucault, Borges, Heterotopia: Producing Knowledge in Other spaces." *Foucault Studies* 9: 54–70.

Tumber, Howard. 2004. "Prisoners of News Values?" In Stuart Allan and Barbie Zelizer (eds.), *Reporting War: Journalism in Wartime*, 190–205. London: Routledge.

Tumber, Howard. 2013. "The Role of the Journalist in Reporting International Conflicts." In Stephen J. A. Ward (ed.), *Global Media Ethics: Problems and Perspectives*, 50–68. Chichester, West Sussex, UK: Wiley-Blackwell.

UNHCR. 2020. "Figures at a Glance." Accessed 30 April 2021. www.unhcr.org.

Van Reybrouck, David. 2014. *Congo: The Epic History of a People*. London: Fourth Estate.

Veis, Greg. 2007. "Baghdad Confidential." In *Mother Jones*, July–August: 13+. Foundation for National Progress.

Vicente, Paulo Nuno. 2013. "The Nairobi Hub: Emerging Patterns of How Foreign Correspondents Frame Citizen Journalists and Social Media." *Ecquid Novi: African Journalism Studies* 34: 36–49.

Vogel, Christoph. 2014. "Islands of Stability or Swamps of Insecurity? MONUSCO's Intervention Brigade and the Danger of Emerging Security Voids in Eastern Congo." *Africa Policy Briefs* 9 (February 2014): 1–10.

Vogel, Christoph, Josaphat Musamba, and Ben Radley. 2018. "A Miner's Canary in Eastern Congo: Formalisation of Artisanal 3T Mining and Precarious Livelihoods in South Kivu." *Extractive Industries and Society* 5: 73–80.

Wacquant, Loïc J. D. 1995. "Pugs at Work: Bodily Capital and Bodily Labour among Professional Boxers." *Body & Society* 1: 65–93.

Wahl-Jorgensen, Karin. 2016. "Emotion and Journalism." In Tamara Witschge, Christopher William Anderson, David Domingo, and Alfred Hermida (eds.), *The Sage Handbook of Digital Journalism*, 128–43. Thousand Oaks, CA: Sage.

Wahl-Jorgenson, Karin, and Mervi Pantti. 2013. "Ethics of Global Disaster Reporting: Journalistic Witnessing and Objectivity." In Stephen J. A. Ward (ed.), *Global Media Ethics: Problems and Perspectives*, 191–213. Chichester, West Sussex, UK: Wiley-Blackwell.

Wainaina, Binyavanga. 2005. "How to Write about Africa." *Granta* 92: 91–95.

Wainaina, Binyavanga. 2012. "How Not to Write about Africa in 2012—a Beginner's Guide." *The Guardian*, June 4, 2012. http://www.theguardian.com/commentisfree/2012/jun/03/how-not-to-write-about-africa.

Wainer, Devorah. 2010. "Beyond the Wire: Levinas vis-à-vis Villawood: A Study of Emmanuel Levinas's Philosophy as an Ethical Foundation for Asylum Seeker Policy." PhD diss., University of Technology Sydney.

Waisbord, Silvio. 2013. *Reinventing Professionalism: Journalism and News in Global Perspective*. Cambridge: Polity.

Wallace, Tina, Lisa Bornstein, and Jennifer Chapman. 2007. *The Aid Chain: Coercion and Commitment in Development NGOs*. Rugby: Practical Action.

Ward, Stephen J. A. 2013. "Introduction: Media Ethics as Global." In Stephen J. A. Ward (ed.), *Global Media Ethics: Problems and Perspectives*, 1–9. Chichester, West Sussex, UK: Wiley-Blackwell.

"Where Have All the Elephants Gone." 2013. *Foreign Correspondent*. Television program, ABC-TV, Sydney. February 5, 2013.

Wilding, Derek, Peter Fray, Sacha Molitorisz, and Elaine McKewon. 2018. "The Impact of Digital Platforms on News and Journalistic Content." Sydney: Centre for Media Transition, University of Technology Sydney.

Williams, Kevin. 2019. *A New History of War Reporting*. New York: Routledge.

Willis, Chris. 2016. "How Well Do We Cover the world?" Panel discussion, Macleay International Reporting Conference.

Wolf, Eric R. 1982. *Europe and the People without History*. Berkeley: University of California Press.

Wright, Kate. 2018. *Who's Reporting Africa Now? Non-governmental Organizations, Journalists, and Multimedia*. New York: Peter Lang.

Wyatt, Wendy N. 2010. "The Ethical Obligations of News Consumers." In Christopher Meyers (ed.), *Journalism Ethics: A Philosophical Approach*, 283–94. New York: Oxford University Press.

Wyschogrod, Edith. 2002. "Language and Alterity in the Thought of Levinas." In Simon Critchley and Robert Bernasconi (eds.), *The Cambridge Companion to Levinas*, 188–205. New York: Cambridge University Press.

Zachary, G. Pascal. 2012. *Hotel Africa: The Politics of Escape*. North Charleston, SC: CreateSpace.

Zelizer, Barbie. 2017. *What Journalism Could Be*. Cambridge: Polity.

Index

For the benefit of digital users, indexed terms that span two pages (e.g., 52–53) may, on occasion, appear on only one of those pages.

Tables and figures are indicated by *t* and *f* following the page number

Printed in the USA/Agawam, MA
November 1, 2022

800593.022